Mechanical Engineering Scie

Mechanical Engineering Science

A. Jackson
Senior Lecturer in Mechanical and
Aeronautical Engineering
The Hatfield Polytechnic

Longman
London and New York

LONGMAN GROUP LIMITED

London and New York

Published in the United States of America by
Longman Inc., New York.

*Associated companies, branches and representatives
throughout the world*

© Longman Group Ltd 1971

All rights reserved. No part of this publication
may be reproduced, stored in a retrieval system,
or transmitted in any form or by any means, electronic,
mechanical, photocopying, recording, or otherwise,
without the prior permission of the Copyright owner.

First published 1971
Second impression 1976
Third impression 1978

ISBN 0 582 42549 2 (paper)

Printed in Great Britain by
Richard Clay (The Chaucer Press) Ltd
Bungay, Suffolk

Preface

This text deals with the topics which must be covered in the Mechanical Engineering Science syllabus at the O1 stage of the Ordinary National Certificate in Engineering, Naval Architecture or Mining. It should also be useful to students taking an Ordinary National Diploma course in Engineering.

The text throughout is written in SI units and the units are presented in squared brackets to separate them from the numerical part of the calculation. It is essential that an engineering student becomes as familiar with units as with numerical calculations.

A considerable number of worked examples are included in the text and no attempt has been made to shorten these solutions. A fault with many examination scripts is the poor presentation of solutions and it is hoped that the layout of the worked examples will assist students to achieve a better presentation of their work.

I wish to express my thanks to Mr P. D. Collins and to my colleagues Mr D. J. Chapple, Mr R. J. Frost, and Mr W. J. Frost for reading various parts of the manuscript and offering helpful criticism and advice.

A. JACKSON

Contents

1. UNITS AND BASIC QUANTITIES — 1
   - 1.1 The SI units — 1
   - 1.2 Definitions of basic quantities — 1
   - 1.3 Force — 2
   - 1.4 Work — 2
   - 1.5 Mass and weight — 3
   - 1.6 Scalar quantity — 4
   - 1.7 Vector quantity — 4
   - 1.8 Addition of vectors — 4

2. STATICS—FORCES — 6
   - 2.1 Statics — 6
   - 2.2 Force — 6
   - 2.3 Equilibrium — 7
   - 2.4 Resultant — 8
   - 2.5 Equilibrant — 8
   - 2.6 Equilibrium of a system of concurrent co-planar forces — 9
   - 2.7 Resolution of forces — 16
   - 2.8 Bow's notation — 19
   - 2.9 Pin-jointed frameworks — 21
   - 2.10 Nature of force in a member — 21

3. MOMENTS — 34
   - 3.1 Moment of a force — 34
   - 3.2 Resultant moment — 35
   - 3.3 Equilibrium of a system of co-planar forces — 35
   - 3.4 Couple — 44
   - 3.5 Centre of mass (centre of gravity) — 47

4. STRESS AND STRAIN — 61
- 4.1 Transmission of forces — 61
- 4.2 Direct stress — 61
- 4.3 Strain — 62
- 4.4 Hooke's law — 63
- 4.5 Modulus of elasticity — 63
- 4.6 Stress–strain graph or load–extension graph — 65
- 4.7 Factor of safety — 66
- 4.8 Tensile strength — 67
- 4.9 Tensile test results — 68
- 4.10 Shear stress — 71

5. FRICTION — 78
- 5.1 Friction — 78
- 5.2 Coefficient of friction — 79
- 5.3 Laws of friction — 80
- 5.4 Angle of friction — 82
- 5.5 Inclined plane—angle of repose — 82
- 5.6 Friction in a journal bearing — 88
- 5.7 Advantages and disadvantages of friction — 90

6. SIMPLE MACHINES — 93
- 6.1 The machine — 93
- 6.2 The ideal machine — 93
- 6.3 Mechanical advantage — 93
- 6.4 Velocity ratio — 94
- 6.5 Efficiency of a machine — 94
- 6.6 Law of a machine — 95
- 6.7 Limiting efficiency of a machine — 95
- 6.8 Overhauling — 97
- 6.9 Examples of simple machines — 98
- 6.10 Simple wheel and axle — 98
- 6.11 Pulley block systems — 100
- 6.12 The screw-jack — 103
- 6.13 The differential wheel and axle — 104
- 6.14 Weston differential pulley block — 106

7. KINEMATICS—VELOCITY AND ACCELERATION — 112
- 7.1 Kinematics — 112
- 7.2 Distance and displacement — 112
- 7.3 Speed — 113

viii Mechanical engineering science

7.4	Velocity	113
7.5	Acceleration	113
7.6	Distance–time graph	114
7.7	Speed–time graph	115
7.8	Constant (or uniform) acceleration	117
7.9	Motion under gravity	124
7.10	Resolution of velocities	124
7.11	Relative velocity	127

8. KINETICS—LAWS OF MOTION   135
8.1	Kinetics	135
8.2	Force and motion	135
8.3	Momentum	135
8.4	Newton's laws of motion	136
8.5	The kinetic equation of motion	136
8.6	Conservation of momentum	144
8.7	Impact of a fluid jet	147

9. ANGULAR MOTION   153
9.1	Angular motion	153
9.2	The radian	153
9.3	Angular velocity	154
9.4	Relationship between angular and linear velocity	154
9.5	Constant angular acceleration	155
9.6	Centripetal acceleration	159
9.7	Centripetal and centrifugal forces	160
9.8	Balancing of co-planar masses	163
9.9	Relative velocity of two points on a rotating link	168

10. MEASUREMENT OF TEMPERATURE AND PRESSURE   175
10.1	Measurement of temperature	175
10.2	Mercury (or alcohol) in glass thermometers	175
10.3	The Beckman thermometer	176
10.4	The gas thermometer	178
10.5	Thermocouples	179
10.6	The radiation pyrometer	180
10.7	The optical pyrometer	181
10.8	The resistance thermometer	181
10.9	Measurement of pressure	182
10.10	The liquid column barometer	183
10.11	The aneroid barometer	184

10.12	The manometer	185
10.13	The inclined manometer	186
10.14	The Bourdon pressure gauge	186
10.15	The engine indicator	187
10.16	Pressure–volume diagram. Work done	189
10.17	The indicator diagram	190

11. GASES — 192

11.1	Nature of a gas	192
11.2	Kinetic theory of gases	192
11.3	Perfect gas or ideal gas concept	193
11.4	Boyle's law	193
11.5	Charles' law	194
11.6	Dalton's law of partial pressures	195
11.7	The characteristic gas equation	195
11.8	The mole (or mol)	200
11.9	Avogadro's hypothesis	200
11.10	Universal gas constant	200
11.11	Volume of one mole of a gas	201
11.12	Specific heat capacities	203
11.13	Relationship between specific heat capacities and gas constant	204
11.14	Conversion of a volumetric analysis of a gas mixture to a mass analysis	207
11.15	Conversion of a mass analysis to a volumetric analysis	208

12. COMBUSTION — 214

12.1	Combustion	214
12.2	Fuels	214
12.3	Atoms and molecules	215
12.4	Combustion of elements	216
12.5	Minimum air required for combustion	217
12.6	Excess air supply	218
12.7	Products of combustion	221
12.8	Calorific value of a fuel	227
12.9	Determination of the calorific value of a fuel	229

13. HEAT TRANSFER — 240

13.1	Modes of heat transfer	240
13.2	Conduction	240
13.3	Convection	240

x Mechanical engineering science

13.4 Radiation 241
13.5 Coefficient of thermal conductivity 241
13.6 Conduction through a material 242
13.7 Conduction through a plane composite wall 243
13.8 Stefan–Boltzman law for radiation 247
Answers to Problems 253
Index 260

Symbols and Abbreviations

SYMBOLS

Term	Symbol
length	l
distance	s
radius	r
diameter	d
area	A
volume	V
time	t
velocity	u, v
angular velocity	ω
acceleration	a
angular acceleration	α
mass	m
density	ρ
force	F
pressure	p
work	W
efficiency	η
direct stress	σ
direct strain	ε
modulus of elasticity	E
temperature, thermodynamic	T
temperature, other than thermodynamic	θ
heat capacity	C
specific heat capacity	c
specific heat capacity at constant pressure	c_p
specific heat capacity at constant volume	c_v
thermal conductivity	k
relative molecular mass	M
number of molecules	N
number of moles	n
radiant energy	W
radiant power	P
radiant intensity	I
radiance	L
radiant emittance	M
irradiance	E

ABBREVIATIONS

Term or Quantity	Abbreviation
metre	m
square metre	m²
cubic metre	m³
litre	l
second (time)	s
minute (time)	min
hour	h
degree, minute, second (angle)	° ′ ″
radian	rad
radian per second	rad/s
revolution per minute	rev/min
gramme	g
newton	N
joule	J
watt	W

1

Units and basic quantities

1.1 The SI units

SI, or to give the full name Système International d'Unités (International System of Units), is the system of units introduced by a general conference of weights and measures in 1960.

SI is based upon six basic quantities:

QUANTITY	UNIT	SYMBOL
Mass	kilogramme	kg
Length	metre	m
Time	second	s
Temperature	kelvin	K
Electric current	ampere	A
Luminous intensity	candela	cd

SI is a coherent (consistent) system of units, i.e. the product or quotient of any two unit quantities is the unit of the resultant quantity. For example, unit force results from multiplying unit mass by unit acceleration.

1.2 Definitions of basic quantities in mechanical engineering science

MASS

The unit of mass, the kilogramme, is the mass of a platinum–iridium cylinder kept at the International Bureau of Weights and Measures at Sèvres, near Paris.

2 Mechanical engineering science

LENGTH

The unit of length, the metre, is the distance between two lines on a platinum–iridium bar which is also kept at the International Bureau of Weights and Measures. The measurement is made at 0°C and standard atmospheric pressure.

TIME

The unit of time, the second, is based upon the mean solar day, this being the average time between successive transits of the sun, i.e.

$$1 \text{ second} = 1/86\,400 \text{ of a mean solar day}$$

Note: The international definitions of the metre and the second are more complex than those given and may be obtained from B.S. 3763.

TEMPERATURE

The unit of thermodynamic temperature, the Kelvin, is the degree interval on the thermodynamic scale. The temperature of the ice point of water on this scale is 273.15 K—corresponding to 0°C on the Celsius scale.

The temperature intervals on the Kelvin and Celsius scales are equal.

1.3 Force

The unit of force in this system is the newton (N). It is a derived unit and is defined as the force which is required to cause a mass of 1 kilogramme to accelerate at a rate of 1 metre/second^2. Thus,

$$1 \text{ N} = 1 \text{ kg m/s}^2$$

1.4 Work

When a force moves through a distance then work is done by the force. The unit of work is the joule (J) which is defined as the work done by a force of 1 newton moving through a distance of 1 metre in the direction of the force. Thus,

$$1 \text{ J} = 1 \text{ N m}$$

This unit is also used for the measurement of energy and quantity of heat, since work and heat are both transient forms of energy.

In the case of a torque:

Work done = torque × angle turned through in radians

In order to differentiate between the units of work and turning moment it is recommended that the joule is used for the measurement of work and energy while the newton-metre is used to refer to the magnitude of a moment or torque. Furthermore, it should be appreciated that work results from a force acting through a distance in its own direction, whereas a moment is obtained by multiplying a force by a distance which is perpendicular to the line of action of the force.

1.5 Mass and weight

The mass of a body is a measure of the quantity of matter within that body. This quantity of matter will be the same irrespective of where the body is located, be it on the surface of the earth or at any other position in the universe. Thus, the mass of a body is a fundamental property of that body.

The Law of Universal Gravitation, propounded by Sir Isaac Newton, states that every body in the universe is attracted to every other body by a force which is proportional to the product of the masses and inversely proportional to the square of their distance apart. This force of attraction is virtually negligible between two bodies on the surface of the earth, but when one of the bodies is the earth, or any other planet, this force of attraction becomes quite considerable and is known as the gravitational force. The particular gravitational force with which a body is attracted towards the earth, the moon, or any other planet, is known as the weight of that body. Now because the earth is not a perfect sphere the weight of a given mass varies over the earth's surface. Furthermore, the weight of a given mass on the surface of the moon is only one-sixth (approximately) of the weight of the same mass on the surface of the earth. Thus, to talk of the weight of a body as a fundamental property of that body is incorrect. With systems of units used prior to the introduction of SI confusion frequently arose because the same unit was used for mass as for weight. With SI a clear distinction is made between units of force and mass and when a force arises in statics as a result of the gravitational action on a mass then the dead load in kilogrammes must be converted into a force in newtons.

Throughout this book the gravitational acceleration is taken as the

4   Mechanical engineering science

standard gravitational acceleration of the earth, i.e. 9.806 65 m/s² (or 9.81 m/s² for calculation purposes). Thus,

Gravitational force in newtons = $9.81m$ (ref. Section 8.5)

where m = mass of body in kilogrammes.
Furthermore, whenever the term weight is used it is understood to imply the gravitational force acting on the mass.

The value of the standard gravitational acceleration was obtained from an experiment carried out at sea level at a latitude of 45°N, i.e. midway between the equator and the north pole. This experiment was carried out during the early part of this century and the resulting value for the acceleration due to gravity was accepted internationally as standard gravitational acceleration. Subsequent work revealed that the value of the gravitational acceleration at this latitude was not exactly 9.806 65 m/s². However, to avoid any confusion this value has been retained as standard. The value of the acceleration due to gravity varies from approximately 9.780 m/s² at the equator to 9.832 m/s² at the poles.

1.6 Scalar quantity

A quantity which possesses magnitude only is known as a scalar quantity, e.g. mass, speed, distance, energy.

1.7 Vector quantity

A vector quantity is one which possesses both magnitude and direction, e.g. force, velocity. A vector quantity can be represented by a line drawn to a certain length and in a certain direction. A vector does not have a position in space.

1.8 Addition of vectors

Suppose *ab* represents a vector quantity, to a certain scale, and *bc* represents a similar vector quantity, to the same scale, then *ac* represents the resultant of *ab* and *ac*, to the same scale. This is shown in Fig. 1.1. Then,

$$ab + bc = ac$$

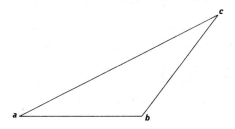

1.1 Vectors

If one vector is to be subtracted from another then it is necessary to change its sign and add it to the other vector, i.e. reverse its direction. Thus,

$$ac - bc = ac + cb = ab$$

Prefixes denoting decimal multiples or sub-multiples

MULTIPLE OF UNIT	PREFIX	SYMBOL
10^{12}	tera	T
10^{9}	giga	G
10^{6}	mega	M
10^{3}	kilo	k
10^{2}	hecto*	h
10^{1}	deca*	da
10^{-1}	deci*	d
10^{-2}	centi*	c
10^{-3}	milli	m
10^{-6}	micro	μ
10^{-9}	nano	n
10^{-12}	pico	p
10^{-15}	femto	f
10^{-18}	atto	a

* Not recommended.

2

Statics—Forces

2.1 Statics

Statics is the branch of mechanics that deals with the forces on bodies which are not accelerating, i.e. they are at rest, or in a state of steady motion.

2.2 Force

A force is defined as that which tends to bring about a change in the motion of a body.

A force has both magnitude and direction and is therefore a vector quantity. As such, it can be represented by a line drawn to scale in a definite direction.

It is very important to appreciate that to every force there is an equal and opposite reacting force. For example, consider a man standing on a wooden plank which is supported by two trestles, as shown in Fig. 2.1.

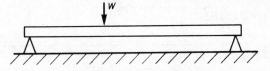

2.1 Load on a plank which is supported on trestles

The weight of the man, i.e. the gravitational force acting on the man, acts downwards on the plank. The man is prevented from falling by the upward force from the plank. We have said that the weight of the man acts downwards on the plank and in turn this is prevented from collapsing by means of the supporting forces from the trestles. Thus, the

Statics—forces 7

forces acting on the plank are shown in Fig. 2.2, the magnitudes of the forces F_1 and F_2 being such that the plank has no tendency to move, i.e. it is in a state of equilibrium.

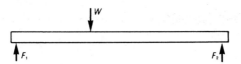

2.2 Forces acting on the plank

Now, if the trestles are pushing upwards on the plank with forces F_1 and F_2 then the forces exerted downwards on the trestles by the plank are also F_1 and F_2, respectively. Likewise, if the trestles are not to sink downwards then the ground must supply upward supporting forces F_1 and F_2 respectively. Thus, the forces acting on the trestles are as shown in Fig. 2.3. The trestles in turn apply forces F_1 and F_2 to the ground.

2.3 Forces acting on the trestles

It should be evident at this stage that it is important to consider the forces which act on a body and not the reacting forces which the body exerts upon other bodies. A diagram showing the forces which act on a body is known as a free body diagram.

2.3 Equilibrium

A body is said to be in a state of equilibrium when the forces which act on it are entirely balanced. If a body is at rest then the effect of such a system of forces is to leave it at rest, while if a body is moving at a constant speed it will continue with the same steady motion.

A body cannot be in equilibrium under the action of a single force, and if two forces act on a body it can only be in equilibrium if the forces are of equal magnitude but act in opposite directions along the same line of action. The conditions necessary for a body to be in equilibrium under the action of more than two forces is dealt with in Sections 2.6 and 3.3.

8 Mechanical engineering science

2.4 Resultant

Suppose a body is acted upon by two non-parallel forces as shown in Fig. 2.4. It is reasonable to suppose that we could find a single force R which would produce the same effect as the two forces F_1 and F_2. Now as force is a vector quantity the magnitude of R can be found by adding F_1 and F_2 vectorially, as shown in Fig. 2.5.

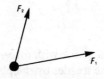
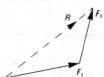

2.4 Two non-parallel forces acting on a body

2.5 Resultant of the two forces

The vectors representing F_1 and F_2 are drawn to scale, in the appropriate directions, their lengths being proportional to the magnitudes of the respective forces.

The force R is said to be the resultant of F_1 and F_2. This procedure can be extended to more than two co-planar forces providing that their lines of action are concurrent, i.e. pass through the same point.

Thus, the resultant of a system of concurrent co-planar forces, which are not in equilibrium, is the single force that can replace the system and produce the same effect.

2.5 Equilibrant

To produce equilibrium of the body which is subjected to the two forces shown in Fig. 2.4 it is necessary to apply a third force to the body such that its line of action passes through the point of intersection of the lines of action of F_1 and F_2. This third force must obviously balance the resultant R. Therefore, this additional third force must be equal to R but opposite in direction along the same line of action.* This force is known as the equilibrant, E. The forces now acting on the body and the force vector diagram are shown in Fig. 2.6 and Fig. 2.7.

It will be seen that the force vector diagram is now a closed triangle (known as a 'triangle of forces') and that the arrows follow in the same direction around the triangle.

---

* If the third force were equal in magnitude to R and opposite in direction but not along the same line of action then the forces acting would constitute a couple (see Section 3.4, page 44).

Statics—forces 9

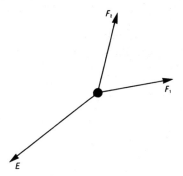

2.6 Body in equilibrium under
the action of three forces

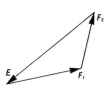

2.7 Force vector diagram

The equilibrant of a system of concurrent co-planar forces is that force which must be added to the system to produce equilibrium.

2.6 Equilibrium of a system of concurrent co-planar forces

From the previous section we know that if a body is in equilibrium under the action of three co-planar forces then they must be concurrent, or a moment will be introduced (see Chapter 3). Furthermore, the forces produce a closed vector diagram.

Let us now consider a body which is acted upon by more than three concurrent co-planar forces. As any two forces can be replaced by a single force we could eventually reduce the system to a three force problem and then draw a force vector diagram to find out if the body is in equilibrium. It is not necessary, however, to draw a series of separate diagrams to determine whether the body is in equilibrium. The whole process can be carried out on a single vector diagram.

Consider a body which is acted upon by the system of forces shown in Fig. 2.8. This diagram is known as a space diagram—it indicates the directions of the forces but the length of line is not proportional to the magnitude of the force. The force vector diagram, Fig. 2.9, is obtained by drawing vectors to scale, and in the appropriate directions, to represent the forces F_1, F_2, etc. It will be seen that the dotted vector R_1 is the resultant of F_1 and F_2, while R_2 is the resultant of R_1 and F_3 or, in other words, the resultant of F_1, F_2, and F_3. If the force vector diagram closes, i.e. ends up at the position from which it was started, then the

10 Mechanical engineering science

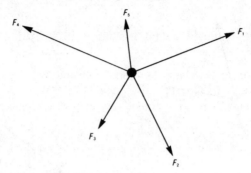

2.8 System of co-planar forces

body is in equilibrium. If this diagram does not close then there is a resultant force acting upon the body.

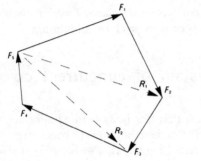

2.9 Force vector diagram for system of co-planar forces

2.10 Alternative force vector diagram

The forces involved in Fig. 2.9 do not have to be drawn in any particular order and Fig. 2.10 shows the same forces drawn in an alternative order. However, it is a good practice for future work to adopt the procedure of working either clockwise or anticlockwise around the forces involved at a particular point.

The force vector diagrams of Fig. 2.9 and Fig. 2.10 are often referred to as a 'polygon of forces'.

To summarise: If a body is in equilibrium under the action of a system of concurrent, co-planar forces, then the force vector diagram must close.

Example 2.1

A drilling machine, of mass 1224 kg, is lifted by an overhead crane using two chains. If the chains make angles of 30° and 50° with the vertical, determine the tension in each chain.

Statics—forces 11

Solution

Gravitational force on machine $= 1224 \times 9.81$
$= 12\,000$ N

Let the tensions in the chains be T_1 and T_2 respectively as shown on the space diagram of Fig. 2.11.

The force vector diagram is drawn, to scale, for equilibrium of the machine and is constructed as follows:

1. Draw vector **ab** vertically to represent the force on the machine of 12 000 N.
2. Because the three forces which act at the junction of the chains are in equilibrium the vector triangle of forces must close. Therefore, lines are drawn from **a** and **b** parallel to the directions of the chains to meet at **c**.

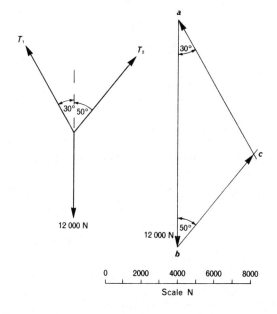

2.11 Space diagram   Force vector diagram
   (Free body diagram for machine)
                     Example 2.1

As the machine is in equilibrium the arrows must follow in the same direction around the vector diagram, so giving the directions of the forces in the chains. With this problem it was, of course, known that

12 Mechanical engineering science

the forces in the chains must be acting upwards on the machine. However, if the directions of the forces had not been known then they would have been obtained from the vector diagram.

Then, vectors *ca* and *bc* represent the tensions in the chains T_1 and T_2 respectively.

From the force vector diagram:

$$bc = 6100 \text{ N} = T_2 \quad \text{and} \quad ca = 9350 \text{ N} = T_1$$

The tensions in the chains are 9350 N and 6100 N.

Example 2.2

Determine the magnitude and direction of the resultant of the coplanar force system shown in Fig. 2.12.

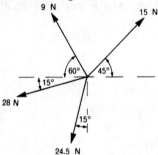

2.12 Space diagram—Example 2.2

Solution

The space diagram for the problem is Fig. 2.12.

The force vector diagram, or polygon of forces, is drawn in Fig. 2.13.

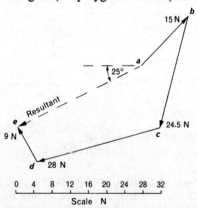

2.13 Force vector diagram—Example 2.2

Statics—forces

This diagram is constructed by drawing vector *ab*, to scale, to represent the force of 15 N; vector *bc*, to scale, to represent the force of 24.5 N, etc.

The resultant is then given by vector *ae*.

From the force polygon this resultant has a magnitude of 30 N and acts at an angle of 25° below the horizontal, to the left.

Example 2.3

A ladder 6 m long, having a mass of 70 kg, leans against a smooth* vertical wall. The ladder is inclined at 65° to the horizontal and its lower end rests on a rough horizontal surface. Assuming that the ladder does not slip determine the reactions at the wall and the ground.

Solution

A scale diagram is given in Fig. 2.14.

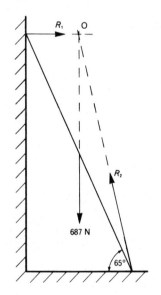

2.14 Space diagram (to scale)　　2.15 Force vector diagram
　　　　　Example 2.3

The gravitational force acting on the ladder (70 × 9.81 = 687 N) is assumed to act at its mid-length. The only other forces acting on the ladder are the reactions from the wall and the ground.

* The term smooth is used to imply a zero coefficient of friction, in which case the reaction must be perpendicular to the surface.

Now, as the wall is smooth, the reaction R_1 must be perpendicular to it. This reaction meets the line of action of the gravitational force at O and therefore, as the ladder is in equilibrium the reaction from the ground R_2, must also pass through O. The reason for drawing Fig. 2.14 to scale should now be apparent—from it we obtain the directions of the forces.

Thus, the directions of all the forces acting on the ladder are now known and the force vector diagram (Fig. 2.15) may be drawn, as follows:

Draw vector *ab* vertically to represent the gravitational force of 687 N. From *a* and *b* draw lines parallel to the reactions R_1 and R_2 to meet at *c*. From Fig. 2.15:

$$R_1 = bc = 160 \text{ N}$$
$$R_2 = ca = 705 \text{ N}$$

The reactions from the wall and the ground are 160 N and 705 N respectively.

Example 2.4

A uniform steel beam AB is 5 m long and has a mass of 320 kg. It is hinged to a vertical wall at A and maintained in a horizontal position by a chain from B to a point C on the wall, which is 6 m vertically above A. Determine the tension in the chain and the magnitude and direction of the reaction at the hinge.

Solution

Gravitational force on beam = $320 \times 9.81 = 3140$ N

As the beam is uniform this force is assumed to act at its mid-length.
A space diagram is drawn to scale, in Fig. 2.16, so that the directions of the forces involved can be obtained.

Let the tension in the chain be T and the reaction at hinge A be R acting at angle θ as shown.

The free body diagram showing the forces acting on the beam is drawn as Fig. 2.17. The line of action of the gravitational force meets the line of action of the tension in the chain (T) at point O. Now if a body is in equilibrium under the action of three co-planar forces then the forces must be concurrent. Therefore, the reaction R must also pass through O. Now that the directions of the three forces acting on the beam are known the force vector diagram can be drawn as in Fig. 2.18.

Statics—forces 15

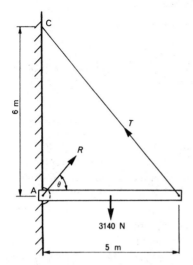
2.16 Space diagram (to scale)

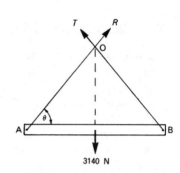
2.17 Free body diagram for beam
Example 2.4

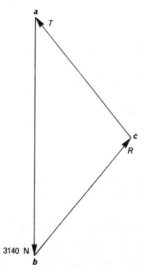
2.18 Force vector diagram—Example 2.4

Vector *ab* is drawn vertically, to scale, to represent the force of 3140 N. From *b* and *a* lines are drawn parallel to the directions of R and T to meet at *c*. Vectors, *bc* and *ca* then represent the forces R and T respectively in both magnitude and direction.

16 Mechanical engineering science

From Fig. 2.18:

Tension in chain = T = ca = 2040 N
Reaction at hinge A = R = bc = 2040 N acting upwards at an angle of 50° to the horizontal.

2.7 Resolution of forces

We have seen that two forces can be combined vectorially to give a single force, the resultant. Conversely, a single force can be replaced by two separate forces whose vector sum is equal to the original force in magnitude and direction. These two separate forces must also pass through the same point of application as the original force. This process is known as resolving a force into components.

Many problems involving forces can often be simplified by resolving the forces into two directions at right angles, the components then being referred to as rectangular components. Referring to Fig. 2.19 the force F can be resolved into rectangular components F_H and F_V, as shown.

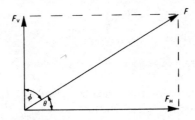

2.19 Resolution of a force

Then,

Component $F_H = F \cos \theta$

and

Component $F_V = F \cos \phi = F \sin \theta$ (since $\theta + \phi = 90°$)

Now, if a body is in equilibrium under the action of a system of concurrent co-planar forces there can be no resultant force on the body. Thus, the resultant of the resolved components in any two directions must each be zero and these two directions are usually taken at right angles for mathematical simplicity. Even when a body is in equilibrium under the action of a system of co-planar forces that are not concurrent there is still no resultant force in any direction, but this is not a sufficient condition for equilibrium (see Section 3.3).

Example 2.5

A casting having a mass of 2200 kg is lifted by an overhead crane using two steel ropes. If the inclination of the ropes to the vertical are 24° and 38° respectively, determine the tension in each rope.

Solution

$$\text{Gravitational force on casting} = 2200 \times 9.81$$
$$= 21\,582 \text{ N}$$

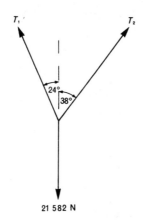

2.20 Space diagram—Example 2.5

Let the tensions in the steel ropes be T_1 and T_2 respectively, as shown on the space diagram, Fig. 2.20.

For equilibrium there can be no resultant force in any direction. Therefore—Resolving vertically:

$$T_1 \cos 24° + T_2 \cos 38° = 21\,582$$
$$0.913\,T_1 + 0.788\,T_2 = 21\,582 \quad (1)$$

Resolving horizontally:

$$T_1 \sin 24° = T_2 \sin 38°$$
$$0.407\,T_1 = 0.616\,T_2$$
$$T_1 = 1.513\,T_2 \quad (2)$$

Substituting for T_1 into equation (1):

$$0.913 \times 1.513\,T_2 + 0.788\,T_2 = 21\,582$$
$$(1.382 + 0.788)T_2 = 21\,582$$
$$T_2 = 9946 \text{ N}$$

18 Mechanical engineering science

Then, in equation (2):

$$T_1 = 1.513 \times 9946$$
$$= 15\,048 \text{ N}$$

The tensions in the chains are 15 048 N and 9946 N respectively.

Example 2.6

Determine the magnitude and direction of the resultant of the co-planar force system shown in Fig. 2.21.

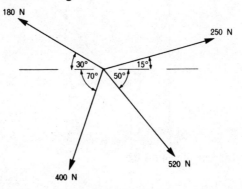

2.21 Space diagram—Example 2.6

Solution

When several forces are involved in a problem such as this a tabular form is preferable.

The sign convention adopted for positive forces is:

$$\text{Horizontally} \rightarrow \qquad \text{Vertically} \uparrow$$

FORCE (N)	HORIZONTAL COMPONENT (N)	VERTICAL COMPONENT (N)
250	$+250 \cos 15° = +241.5$	$+250 \sin 15° = + 64.7$
180	$-180 \cos 30° = -155.9$	$+180 \sin 30° = + 90.0$
400	$-400 \cos 70° = -136.8$	$-400 \sin 70° = -375.9$
520	$+520 \cos 50° = +334.2$	$-520 \sin 50° = -398.3$
	$+283.0$	-619.5

The resultant vertical and horizontal components are shown in Fig. 2.22.

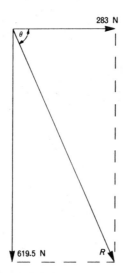

2.22 Resultant forces—Example 2.6

The resultant force R is then given by

$$R^2 = 283^2 + 619.5^2$$

giving

$$R = 680 \text{ N}$$

and

$$\tan \theta = \frac{619.5}{283} = 2.189$$

$$\theta = 65° 27'$$

The resultant of the given force system is 680 N acting at an angle of 65° 27′ below the horizontal, to the right.

It is suggested that, as a useful check, the student should now solve Worked Examples 2.1 and 2.2 by the resolution method.

2.8 Bow's notation

It is often desirable to have a systematic method of referring to the forces acting on a body. This is the purpose of Bow's notation where the spaces between forces are assigned capital letters and the corresponding small letters are used to label the forces on the vector diagram.

Thus in Fig. 2.23 force F_1 is denoted AB on the space diagram and **ab** on the vector diagram. Similarly for the other forces.

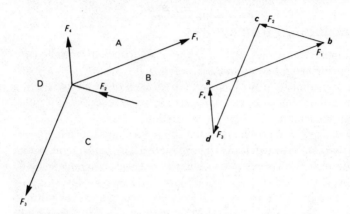

2.23 Space diagram   Vector diagram
Bow's notation

The force vector diagram has been drawn by taking the forces in a clockwise cyclic order around the joint. This clockwise convention has been adopted throughout this book.

The adoption of this convention together with the introduction of Bow's notation enables the arrows to be omitted from the force vector diagram. The significance of this will be appreciated when dealing with pin-jointed frameworks in Section 2.9.

Let us now look at the force diagram of Fig. 2.23 with the arrows and F_1, F_2, etc. removed, as in Fig. 2.24.

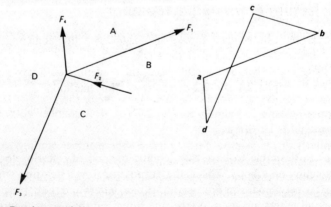

2.24 Bow's notation

Then, between space A and space B we have force *ab* and its direction is from *a* to *b* on the vector diagram. This direction corresponds with that on the space diagram. Similarly for the other forces.

2.9 Pin-jointed frameworks

A pin-jointed framework is a structure comprised of straight rigid members connected by pin-joints. The external loads on such a structure are only applied at the joints and, by reason of its pinned ends, a member of the framework can only carry an axial load, i.e. in a direction along its length.

When dealing with problems on pin-jointed frameworks a force vector diagram can be drawn for each joint in turn, since each joint as well as the complete framework is in equilibrium. A start can only be made at a joint where there are not more than two unknown forces acting and it is frequently necessary to calculate support reactions prior to drawing the force vector diagram. The support reactions can be obtained by taking consideration of moment equilibrium (ref. Chapter 3).

It will be seen that the separate force vector diagrams for each joint can be combined into a single diagram, known as a Maxwell diagram. No arrowheads are inserted on this diagram because of the opposing directions of the forces which a member exerts at its two ends.

The method given in this chapter for determining forces in the members of a framework can only be applied to a pin-jointed framework which is loaded at the joints. The framework must also be statically determinate, i.e. the forces in the framework members can be determined by applying the conditions of statical equilibrium.

2.10 Nature of force in a member

The external forces applied to a member must be balanced by internal resisting forces created within the member if equilibrium is to be achieved.

When the external forces pull on the framework member, so tending to increase its length, the internal forces created in the member act in an opposing sense, as shown in Fig. 2.25(*a*). The member is then in TENSION and is known as a TIE.

Similarly, if the external forces push on the member tending to decrease its length then internal forces are set up as shown in Fig. 2.25(*b*). The member is then in COMPRESSION and is known as a STRUT.

The member forces obtained from the force vector diagram are the internal forces in the member. Thus,

22 Mechanical engineering science

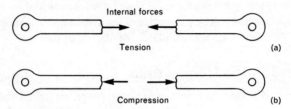

2.25 Nature of force in a member

A member is a TIE when the internal forces pull on a joint.
A member is a STRUT when the internal forces push on a joint.

Example 2.7

Determine the forces in the members of the pin-jointed framework shown in Fig. 2.26(*a*) when it is subjected to the loading shown. What are the reactions at the supports?

Solution

The spaces between the members and the forces are lettered according to Bow's notation. The force vector diagrams, which are constructed to scale, are obtained as follows:

The only joint with a minimum of two unknown forces is joint ABD. Therefore, we must start at this joint.

For joint ABD (Fig. 2.26(*b*)):

1. Draw vector *ab* to scale to represent the force AB of 5 kN.
2. Draw a force through *b* parallel to BD to meet a force drawn through *a* parallel to DA at *d*.
3. Transfer the directions of the forces at this joint, which are governed by the direction of *ab*, onto the space diagram, Fig. 2.26(*a*).

Note the order in which the forces are drawn, i.e. AB, BD, DA, or clockwise around the joint.

At this stage we can proceed to either of the two remaining joints as at each joint only two unknown forces are involved.

Then, for joint DBC (Fig. 2.26(*c*)):
We already know the force in BD from figure (*b*).

1. Draw vector *db* equal in length and parallel to vector *bd*.
2. Draw a line vertically from *b* parallel to the support reaction BC.

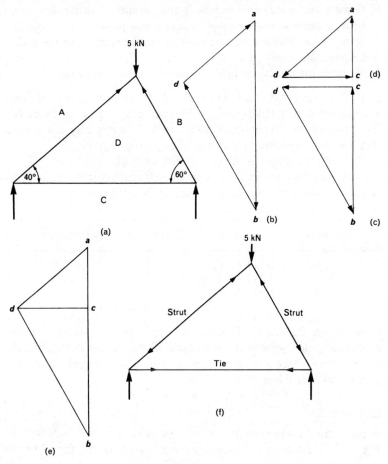

2.26 Pin jointed framework—Example 2.7

3. Draw a line through *d* parallel to force CD to obtain point *c*.
4. Transfer the directions of the forces from vector diagram *bcd* to joint BCD.

These directions are governed by:

(*a*) The force exerted by member DB on joint BCD must be opposite to the force exerted by the same member on joint ABD.
(*b*) The support reaction BC must be upwards.

Joint ADC (Fig. 2.26(*d*)):

We already know the forces in AD, from figure (*b*), and DC, from figure (*c*).

24 Mechanical engineering science

1. Draw a vector *ad* equal in length and parallel to vector *da*.
2. From *d* draw vector *dc* equal in length and parallel to vector *cd*.
3. Joint *c* to *a*. This must also be parallel to the support reaction CA, i.e. vertical.
4. Transfer the directions of the forces onto the space diagram.

It will be seen that figures (*b*), (*c*), and (*d*) can be combined to form a single force vector diagram, or Maxwell diagram, as in figure (*e*). However, no arrows can be sensibly inserted on this diagram as each vector would have two arrows on it in opposite directions.

The complete diagram of force directions is given in figure (*f*).

From this diagram and the separate vector diagrams, or the combined force diagram, the following results are obtained.

MEMBER	FORCE	REACTION	MAGNITUDE
AD	2.54 kN (strut)	CA	1.63 kN
BD	3.89 kN (strut)	BC	3.37 kN
CD	1.95 kN (tie)		

Note: When drawing the force vector diagrams it is essential to proceed around each joint in the same cyclic order, i.e. clockwise throughout this text. Once started the order cannot be changed during the solution of a problem.

Example 2.8

A pin-jointed framework is pinned to a vertical wall as indicated in Fig. 2.27. Determine the magnitude and nature of the forces in the

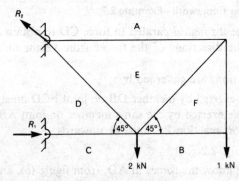

2.27 Framework pinned to a wall—Example 2.8

framework members when they are subjected to the loading shown. What are the magnitudes and directions of the reactions at the wall joints?

Solution

Let the reactions acting on the frame at the wall joints be R_1 and R_2, respectively, as shown in Fig. 2.27. At the lower joint there is one member attached and therefore the line of action of R_1 must be in line with the member. At the upper joint there are two members and therefore the line of action of the reaction R_2 is unknown at this stage.

The framework space diagram is lettered according to Bow's notation as shown. It is important to realise that the wall is not a member and that the space D merely separates the reactions, R_1 and R_2, at the wall joints.

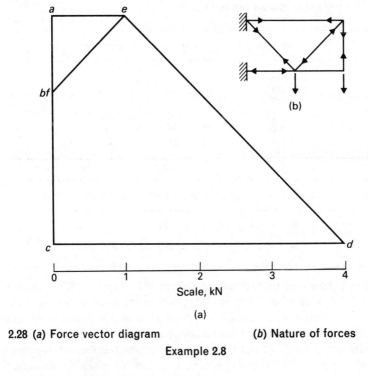

2.28 (a) Force vector diagram (b) Nature of forces
Example 2.8

The force vector diagram, constructed to scale in Fig. 2.28(a), is obtained by:

26 Mechanical engineering science

1. Starting at joint ABF, where there are only two unknown forces, and constructing the force triangle *abf*. In this instance *b* and *f* must be the same point or otherwise we would not return to *a* when drawing vector *fa*. Transfer the force directions to the space diagram.
2. Proceed to joint AFE and obtain force triangle *afe*.
3. Proceed to joint EFBCD.

   Points *e*, *f* and *b* are already fixed.

   Draw *bc*, to scale, to represent the vertical load of 2 kN.

   From *c* draw a line parallel to CD to meet a line drawn from *e* parallel to DE at *d*.

The force vector diagram is now complete and the directions of the forces at the joints are shown in Fig. 2.28(*b*).

The reaction at the lower joint is given by vector *cd* on the force diagram, and the reaction at the top joint, R_2, is given by vector *da*, i.e. R_2 is the equilibrant of forces *ae* and *ed*.

The following results are obtained:

MEMBER	FORCE	NATURE
AE	1 kN	tie
AF	1 kN	tie
BF	0	
CD	4 kN	strut
FE	1.41 kN	strut
DE	4.24 kN	tie

REACTION	MAGNITUDE AND DIRECTION
R_1	4 kN →
R_2	5 kN 37°

Let us now look more closely at member BF. At joint ABF the member AF is vertical and the load applied to the joint is also vertical. Hence, referring to Fig. 2.29, it will be seen that if member BF did carry a force then it would not be possible to achieve horizontal equilibrium at the joint. Thus, for the loading shown, member BF is superfluous as far as load carrying is concerned but this situation no longer exists if the load at joint ABF is applied in any direction other than vertically.

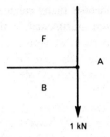

2.29 Joint ABF—Example 2.8

This leads to an important result. If three members meet at a joint and any two of the members are collinear then the third member carries no force. Thus, referring to Fig. 2.30, we can state, without

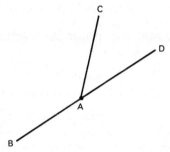

2.30 Equilibrium of joint A

reference to any remaining structure, that the force in AC is zero. If this were not so then equilibrium perpendicular to BAD could not be achieved. However, if a force were applied at joint A having a line of action other than along DAB then the member CA would carry a force. Hence, because a member has zero force in it for a particular loading system does not imply that it will always carry no force.

Problems

1. A crate and its contents have a mass of 600 kg and during a loading process the crate is lifted by two chains which make angles of 20° and 35° with the vertical. Determine the tensions in the chains.

2. The lifting bucket of a dragline is suspended from the jib by two chains. When the mass of the bucket and its ore content is 42 Mg what will be the tensions in the supporting chains if they make angles of 12° and 25° to the vertical?

3. Determine either graphically or by calculation the resultant of the co-planar concurrent system of forces shown in Fig. 2.31.

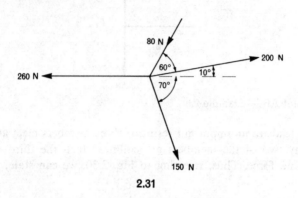

2.31

4. The following system of co-planar forces acts on a body such that their lines of action are concurrent.

80 N at 5°;   30 N at 90°;   35 N at 130°;
70 N at 245°;   120 N at 320°.

The angles are measured anticlockwise from a horizontal left to right datum direction.

Determine by calculation the magnitude and direction of the resultant of this system of forces.

5. A body is maintained in a state of equilibrium by the action of four concurrent co-planar forces. The forces have magnitudes of 70 N, 50 N, 120 N, and 80 N respectively. If the angle between the lines of action of the forces of 70 N and 50 N is 70°, determine the directions in which the other two forces act relative to the force of 70 N.

6. A uniform rod BC having a mass of 8 kg is pivoted at D and maintained in the position shown in Fig. 2.32 by means of the cord AB. If BC = 2 m and BD = 0.75 m determine:

(*a*) the tension in the cord AB;
(*b*) the magnitude and direction of the reaction at the pivot D.

7. A steel beam of length 6 m and having a mass of 200 kg rests with its lower end on a rough horizontal concrete surface while its upper end rests against a smooth vertical wall. When the beam is inclined at 55° to the horizontal, calculate the magnitude and direction of the reactions at the wall and the ground.

Statics—forces 29

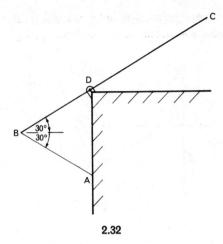

2.32

8. A roller having a mass of 40 kg is maintained in equilibrium on a smooth plane, which is inclined at 25° to the horizontal, by the application of force P_1 horizontally and P_2 parallel to the plane. If $P_2 = 3P_1$ determine the values of P_1 and P_2 and the normal reaction between the plane and the roller.

9. A uniform beam AB is hinged to a wall at A so that it is free to turn in a vertical plane. The beam has a mass of 80 kg and is maintained in a horizontal position by a strut BC such that the angle ABC is 50°. The strut is hinged to the beam at B and to the wall at C, vertically below A. Determine the force in the strut and the magnitude and direction of the reaction at A.

Determine for Problems 10–20 the magnitude and nature of the forces in the members of the pin-jointed frameworks shown. Find also, the magnitude and direction of any support reactions.

10.

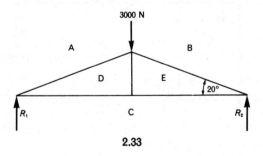

2.33

30 Mechanical engineering science

11.

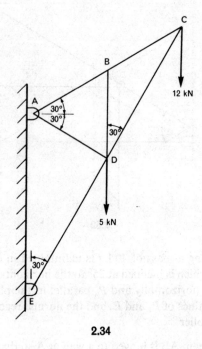

2.34

12.

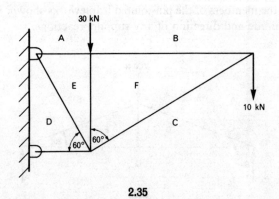

2.35

13.

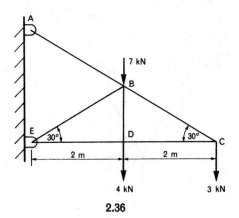

2.36

14.

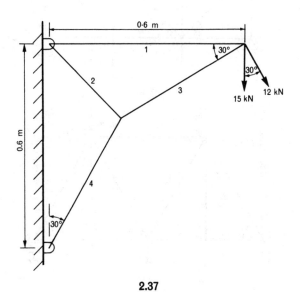

2.37

32 Mechanical engineering science

15.

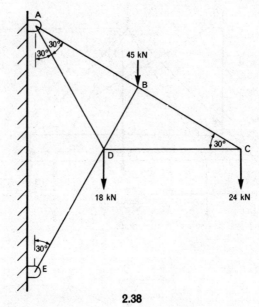

2.38

16.

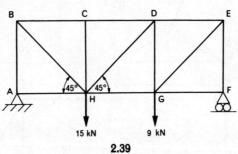

2.39

17.

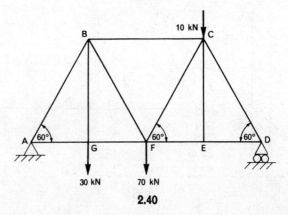

2.40

Statics—forces   33

18.

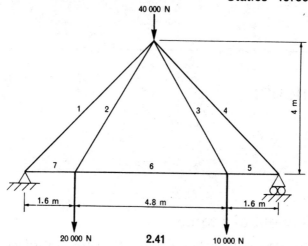

2.41

19.

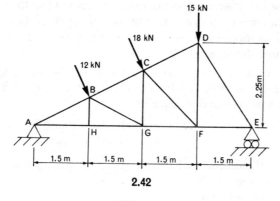

2.42

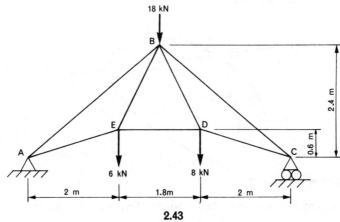

2.43

3

Moments

3.1 Moment of a force

If a body, which is free to turn about a certain axis, is acted upon by a force whose line of action does not pass through that axis then the body will tend to turn. If the distance of the line of action of the force from the axis of rotation is increased then the turning effect, or moment, of the force will also increase.

The moment of a force about an axis is dependant upon the magnitude of the force and the perpendicular distance of its line of action from the axis and is defined by:

Moment of force = force × perpendicular distance from axis to
(about an axis)   line of action of force

The units of moment are therefore force × length, the basic quantity being newton × metre (N m).

Then, referring to Fig. 3.1:

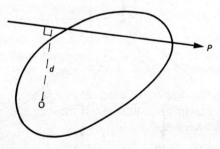

3.1 Moment of a force

Moment of force about axis through $O = P \times d$ (clockwise)

Although the moment of a force refers specifically to a particular axis, when dealing with a system of co-planar forces it is in order to refer to the moment about a point (meaning the axis through that point which is perpendicular to the plane). Thus, with reference to Fig. 3.1, if we say 'moment of force about point O' we really mean 'about the axis through O perpendicular to the plane of the paper'.

3.2 Resultant moment

When a body is acted upon by several forces the resultant moment about any axis is the algebraic sum of the moments of the individual forces about that axis, account being taken of whether the force produces a clockwise or anticlockwise moment about the axis.

3.3 Equilibrium of a system of co-planar forces

We have seen in Chapter 2, Section 2.6, that when a body is in equilibrium under the action of a system of co-planar concurrent forces then there is no resultant force acting on the body. Let us now suppose that the forces are no longer concurrent. This means that some of the forces will exert a moment about any chosen axis and therefore, in addition to there being no resultant force on the body, we shall require a further condition to establish equilibrium. This is that there is also no resultant moment about any axis of the body, or otherwise the body would rotate about that axis with an increasing speed.

This is embodied in the Principle of Moments, which states: If a body is in equilibrium under the action of a system of forces the resultant moment about any axis is zero, i.e. clockwise moments = anticlockwise moments.

Example 3.1

A light uniform beam AB is 8 m long and is simply supported at A and B. The beam carries loads of 260 N at a position 1.8 m from A and 480 N at a position 3.2 m from B. Calculate:

(a) the value of the support reactions at A and B;
(b) the position at which a load of 500 N must be applied to make the support reactions equal.

Solution

(a) A diagram is given in Fig. 3.2.

36 Mechanical engineering science

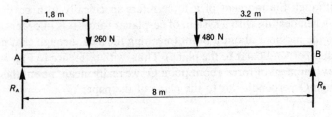

3.2 Simply supported beam—Example 3.1

Let the support reactions at A and B be R_A and R_B respectively as shown. As the beam is in equilibrium there can be

(1) no resultant force on the beam,

$$\therefore R_A + R_B = 260 + 480$$
$$= 740 \text{ N} \qquad (1)$$

(2) no resultant moment about any axis of the beam.

By taking moments of forces about (or through) A, R_A will have no moment about A. Thus the moment equation will involve only one unknown, viz. R_B.
Thus,

Clockwise moments about A = counter-clockwise moments about A

$$260 \times 1.8 + 480 \times 4.8 \, [\text{N m}] = R_B \times 8 \, [\text{m}]$$

$$468 + 2304 = 8 R_B$$

$$R_B = \frac{2772}{8} \frac{[\text{N m}]}{[\text{m}]}$$

$$= 346.5 \text{ N}$$

From equation (1)

$$R_A + R_B = 740 \text{ N}$$

$$\therefore R_A = 740 - 346.5$$

$$= 393.5 \text{ N}$$

The support reactions at A and B are 393.5 N and 346.5 N respectively.
(*b*) The diagram for this part is Fig. 3.3.
Let the position at which a load of 500 N must be applied to make the reactions equal be x metres from A.

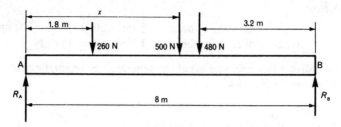

3.3 Example 3.1

For equilibrium there must be:

(1) No resultant force on the beam, i.e.

$$R_A + R_B = 260 + 500 + 480 \text{ N}$$
$$= 1240 \text{ N}$$

But, $R_A = R_B$

$\therefore R_A = 620 \text{ N} = R_B$

(2) No resultant moment about any axis of the beam.
Taking moments about A;

Clockwise moments about A = counter-clockwise moments about A

$$260 \times 1.8 + 500 \times x + 480 \times 4.8 = R_B \times 8$$
$$468 + 500x + 2304 = 4960$$
$$500x = 2188$$
$$x = 4.376 \text{ m}$$

The load of 500 N must be applied at a distance of 4.376 m from A in order to make the reactions equal.

Example 3.2

A uniform beam AB is 6 m long and has a mass of 20 kg/m. It is hinged to a vertical wall at A and carries a load of 600 N at a position 4.2 m from A. The beam is maintained in a horizontal position by means of a steel rope attached to the beam at B and connected to a wall coupling at C which is vertically above A. If the angle ABC is 50° determine:

(a) the tension in the steel rope;
(b) the magnitude and direction of the reaction at the hinge A.

Solution

$$\text{Total mass of beam} = 6 \, [m] \times 20 \, [kg/m] = 120 \, kg$$
$$\therefore \text{Gravitational force on beam} = 120 \times 9.81 = 1177 \, N$$

This force will effectively act at the centre of the beam, i.e. 3 m from A, as shown in Fig. 3.4.

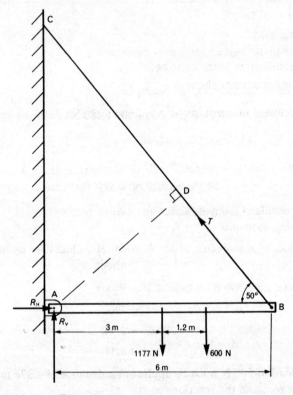

3.4 Hinged beam—Example 3.2

Let the tension in the steel rope be T and the horizontal and vertical components of the reaction at A acting on the beam be R_H and R_V respectively, as shown. Thus the forces shown on Fig. 3.5 are those acting on the beam, i.e. the diagram is a free body diagram for the beam.

Now, as the beam is in equilibrium there is no resultant moment acting on it and by taking moments about A the only unknown force involved is T, since R_H and R_V both act through A.

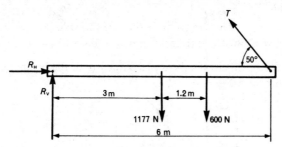

3.5 Free body diagram for beam—Example 3.2

Taking moments about A:

Clockwise moment about A = anticlockwise moment about A

$$1177 \times 3 + 600 \times 4.2 = T \times AD$$

where AD = perpendicular distance from hinge A to the line of action of the force T

$$= 6 \sin 50°$$
$$= 4.596 \text{ m}$$
$$\therefore 3531 + 2520 = 4.596T$$
$$T = \frac{6051}{4.596} = 1317 \text{ N}$$

There must also be no resultant force on the beam in any direction. Therefore, resolving horizontally:

$$R_H = T \cos 50°$$
$$= 1317 \cos 50° = 846 \text{ N}$$

and, resolving vertically:

$$R_V + T \sin 50° = 1177 + 600$$
$$R_V + 1009 = 1777$$
$$R_V = 768 \text{ N}$$

These components are shown on Fig. 3.6, where R is the resultant reaction at A acting at the angle θ shown.

40 Mechanical engineering science

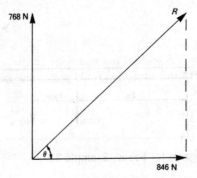

3.6 Resultant forces on beam at hinge A—Example 3.2

Then
$$R^2 = 768^2 + 846^2$$
$$R = 1143 \text{ N}$$
$$\tan \theta = \frac{768}{846} = 0.9078$$
$$\theta = 42.2°$$

The tension in the steel rope is 1317 N and the reaction at hinge A is 1143 N acting upwards at 42.2° to the beam. The above solution should be compared with that of Example 2.4 (p. 14).

Example 3.3

Fig. 3.7 shows two straight members XY and YZ that are pinned at their ends and which lie in a vertical plane. For the applied loading shown, determine the magnitude of the reactions at joints X and Z.

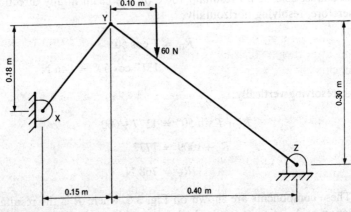

3.7 Example 3.3

Solution

ANALYTICALLY

Consider member YZ and let the reactions at the joints Y and Z be V_Y, H_Y, V_Z, and H_Z respectively, as shown on the free body diagram, Fig. 3.8.

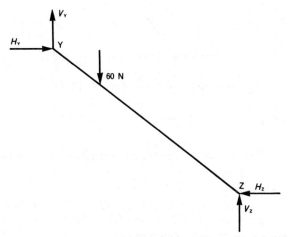

3.8 Free body diagram for member YZ—Example 3.3

For horizontal equilibrium of link YZ

$$H_Y = H_Z \tag{1}$$

And for vertical equilibrium of link YZ

$$V_Y + V_Z = 60 \text{ N} \tag{2}$$

As the joints at Y and Z are pin joints there can be no resultant moment at them if the link is to be in equilibrium. Therefore taking moments about Z:

Clockwise moments about Z = counter-clockwise moments about Z

$$V_Y \times 0.40 + H_Y \times 0.30 = 60 \times 0.30 \tag{3}$$

Considering now member XY, it is evident that as there are no applied forces acting on this member then the reactions at X and Y must be equal and opposite. Furthermore, the reactions at Y acting on member YX must be equal and opposite to those acting on member YZ if equilibrium is to be satisfied at the joint. Thus the reactions on XY are as shown in Fig. 3.9.

42 Mechanical engineering science

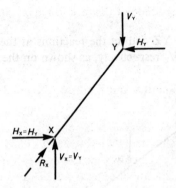

3.9 Free body diagram for member XY—Example 3.2

Again, as link XY is in equilibrium, there can be no resultant moment about X or Y.

Taking moments about X:

$$V_Y \times 0.15 = H_Y \times 0.18 \tag{4}$$

Substituting for H_Y into equation (3)

$$0.40 V_Y + \frac{0.15}{0.18} \times 0.30 V_Y = 60 \times 0.30$$

$$0.65 V_Y = 18$$

$$V_Y = 27.7 \text{ N} = V_X$$

From (4) $\quad H_Y = \dfrac{0.15}{0.18} \times 27.7 = 23.1 \text{ N} = H_X$

From (1) $\quad H_Z = H_Y = 23.1 \text{ N}$

From (2) $\quad V_Z = 60 - V_Y = 60 - 27.7$

$$= 32.3 \text{ N}$$

Joint X:

Let R_X be the magnitude of the reaction at X. Then

$$R_X^2 = V_X^2 + H_X^2$$

$$= 27.7^2 + 23.1^2$$

$$R_X = 36.1 \text{ N}$$

Joint Z:

Let R_Z be the magnitude of the reaction at Z. Then
$$R_Z^2 = V_Z^2 + H_Z^2$$
$$= 32.3^2 + 23.1^2$$
$$R_Z = 39.7 \text{ N}$$

The reactions at joints X and Z are 36.1 N and 39.7 N respectively.

GRAPHICALLY

A solution to this problem can be quickly obtained by using the principles set out in Chapter 2. We saw that if a body was in equilibrium under the action of three co-planar forces then those forces must be concurrent. This is the situation in this problem and the three forces which are acting are:
1. The downward applied force of 60 N
2. The reaction at X.
3. The reaction at Z.

Now at X the reaction must be in the direction of link XY since there are no external forces applied to the member XY. The line of action of this reaction at X meets the line of action of the applied force of 60 N at O, as shown on Fig. 3.10. As the members XY and YZ are in equilibrium it follows that the reaction at Z must also pass through O.

Thus, at this stage, the directions of all the forces are known and it only remains to draw the force vector diagram, Fig. 3.11, to obtain the magnitudes of the reactions.

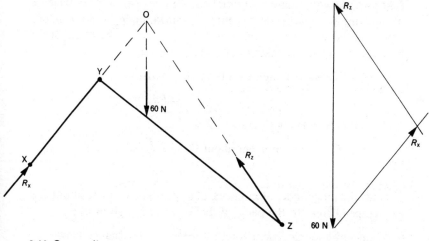

3.10 Space diagram　　　　　　　　3.11 Force vector diagram

Example 3.2

From the force vector diagram the reactions at the joints X and Z are 36 N and 40 N respectively.

3.4 Couple

Consider the case of a body which is acted upon by two parallel forces which are equal in magnitude but opposite in direction, as shown in Fig. 3.12.

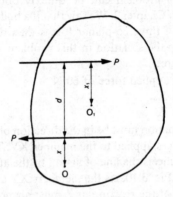

3.12 A couple

Although there is no resultant force on the body, there is a resultant moment which would cause the body to rotate. This resultant moment is known as a couple and equilibrium of the body can only be achieved by applying an equal opposing couple. The magnitude, or moment, of a couple is the same about any axis perpendicular to the plane of the forces and is given by taking moments about O:

$$\text{Clockwise moment about O} = P(d+x) - Px$$
$$= Pd$$

or, taking moments about O_1:

$$\text{Clockwise moment about } O_1 = Px_1 + P(d-x_1)$$
$$= Pd$$

Thus, the magnitude, or moment, of a couple is the same about any axis perpendicular to the plane of the forces and is given by:

Moment of couple = one force × perpendicular distance between their lines of action

Moments 45

Thus, the couple shown in Fig. 3.12 has a magnitude of $P \times d$. The basic unit of a couple is therefore the newton metre (N m).

When a couple is applied to produce rotation of a shaft it is generally referred to as a torque. The tightening of a wing nut, or clock spring, are the simplest examples of the application of two equal and opposite parallel forces to produce a torque, and hence rotation. Frequently, however, rotation results from an external force being reacted by an equal and opposite force which is not collinear. Suppose a force P is applied to the rim of a pulley whose shaft is supported in bearings. There must be a reacting force, equal to P, from the bearings which acts on the shaft, as shown dotted in Fig. 3.13. Thus the pulley is subjected to a torque which produces rotation about its shaft axis.

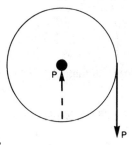

3.13 Forces on a pulley

Let us now consider the tightening of a nut by means of a spanner where a force P is applied at a distance d from the nut, as shown in Fig. 3.14(i). For equilibrium of the spanner a reacting force equal to P must be exerted on the spanner by the nut. This means that the spanner is subjected to a torque of $P \times d$ and will therefore tend to rotate. The nut is therefore subjected to a force P, which is opposite to the reaction of the nut on the spanner, together with a torque of magnitude Pd, and to maintain force equilibrium of the nut there must be a reaction force P from the screw thread.

As the nut is tightened so the resisting torque T acting on the nut will gradually increase. This torque arises from frictional effects at the thread and between the nut and the surface against which it is being tightened. In turn, the nut will exert an opposing torque T on the spanner. The applied forces and torque are shown solid in figures (ii) and (iii), the reacting forces and torque being dotted. Providing that the applied torque Pd is greater than the resisting torque T then further tightening of the nut will occur. The limit is reached when

46 Mechanical engineering science

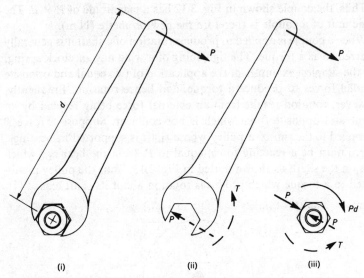

3.14 Forces involved when a nut is tightened by means of a spanner

$Pd = T$, the nut and spanner being in equilibrium with regard to both force and torque.

Example 3.4
A shaft is driven by a belt which passes over a pulley of 200 mm diameter. If the ratio of the belt tensions on the tight and slack sides is 8 : 1 and a net torque of 60 N m is transmitted to the shaft, what will be the tensions in the two sides of the belt?

Solution
Let the tension on the slack side of the belt be P. Then the tension on the tight side will be $8P$. These are shown by solid lines on Fig. 3.15.

As there can be no resultant force on the pulley, since it only rotates, the shaft must apply reaction forces parallel to the tensions in the belt. These are shown dotted in the diagram.

The pulley is therefore subjected to two pairs of equal and parallel forces, i.e. to two torques. As these torques oppose each other the resultant torque transmitted (T) is given by:

$$T = (8P - P)r$$

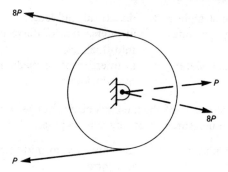

3.15 Example 3.4

where r = radius of pulley = 100 mm = 0.1 m,

$$\therefore 60 \, [\text{N m}] = (8P - P) \times 0.1 \, [\text{m}]$$

$$P = \frac{60}{7 \times 0.1} \frac{[\text{N m}]}{[\text{m}]} = 85.7 \, \text{N}$$

Tension in slack side of belt = 85.7 N

Tension in tight side of belt = 8 × 85.7 = 685.6 N

3.5 Centre of mass (centre of gravity)

It was mentioned in Section 1.5 that when a body is in a gravitational field it is acted upon by the gravitational pull of that field. Thus, all bodies on the earth are acted upon by the gravitational pull of the earth. If a body is divided up into a number of sections then the result of the gravitational pull would be a series of virtually parallel forces acting on the individual sections. The line of action of the resultant of these forces would then lie in a vertical plane and pass through a point known as the centre of mass of the body. Thus, all the mass of the body can be imagined to be concentrated at this position, i.e. at the centre of mass, and the gravitational force on the body will therefore always act through this centre of mass, irrespective of the position of the body. It follows that in order to support a body in a particular position the resultant upward force must also pass through the centre of mass of the body.

The position of the centre of mass of many bodies is easily determined, being coincident with the geometrical centre, and the following list assumes uniform thickness and density of material.

48 Mechanical engineering science

Uniform straight rod: on axis at mid-length
Rectangular plate: at intersection of diagonals and at mid-thickness
Triangular plate: at intersection of medians and at mid-thickness

Note: the intersection of the medians occurs at one third of the height of the triangle measured from any side as base.

Circular plate: at centre and at mid-thickness
Sphere: at centre
Parallelogram plate: at centre and at mid-thickness

In the case of an unsymmetrical body the centre of mass is found by dividing the body up into sections whose individual centre of mass is known and employing the principle of moments to obtain the position of the centre of mass of the complete body.

Suppose a body of mass m has its centre of mass at G, having co-ordinates $\bar{x}$, $\bar{y}$, and $\bar{z}$ respectively. This is illustrated in Fig. 3.16 which

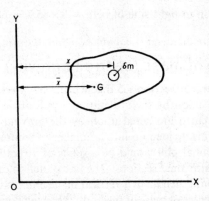

3.16 Centre of mass of a body

shows a view looking in the direction of the Z axis, i.e. the Z axis is through O, perpendicular to the plane of the paper. Consider an elemental mass δm of the body, as shown, having coordinates x, y, and z respectively.

Moment of elemental mass about axis OZ $= x \cdot \delta m$

Total moment of body about axis OZ = sum of moments of all such elemental masses $= \sum x \cdot \delta m$

Total mass of the body,

$$m = \sum \delta m$$

But the total moment of the body about axis OZ $= m\bar{x}$

$$\therefore \bar{x} = \frac{\sum x \cdot \delta m}{m} = \frac{\sum x \cdot \delta m}{\sum \delta m}$$

Similarly,

$$\bar{y} = \frac{\sum y \cdot \delta m}{\sum \delta m} \quad \text{and} \quad \bar{z} = \frac{\sum z \cdot \delta m}{\sum \delta m}$$

When dealing with an area or volume the term centroid is used. The coordinates of a centroid are defined by

$$\bar{x} = \frac{\sum x \cdot \delta A}{\sum \delta A} \quad \text{etc. for an area}$$

or

$$\bar{x} = \frac{\sum x \cdot \delta V}{\sum \delta V} \quad \text{etc. for a volume}$$

Example 3.5
A hole of 80 mm diameter is cut out of a thin uniform rectangular plate PQRS whose dimensions are PQ = RS = 400 mm and QR = PS = 250 mm. If the centre of the cut-out is 80 mm from PQ and 120 mm from PS determine the position of the centre of mass of the remaining plate.

Solution
A diagram is given in Fig. 3.17.
Let the centre of mass of the plate be at G, distance $\bar{x}$ from PS and $\bar{y}$ from PQ, as shown. As the plate is of uniform thickness its mass will be proportional to the area.
Let the mass of the plate be m kilogrammes per square millimetre.

$$\text{Total mass of original plate} = 400 \times 250 \times m$$
$$= 100\,000m \text{ kg}$$
$$\text{Mass of circular plate removed} = \frac{\pi}{4} \times 80^2 \times m$$
$$= 5026m \text{ kg}$$
$$\therefore \text{Mass of plate remaining} = 94\,974m \text{ kg}$$

50 Mechanical engineering science

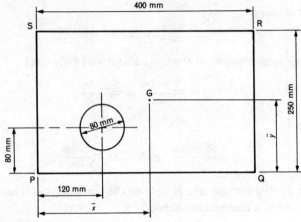

3.17 Example 3.5

Taking moments about axis PQ:

Moment of plate remaining about PQ + moment of cut-out plate about PQ = moment of original plate about PQ

$$94\,974m\bar{y} + 5026m \times 80 = 100\,000m \times 125$$

$$\bar{y} = 127.4 \text{ mm}$$

Similarly, taking moments about axis PS:

$$94\,974m\bar{x} + 5026m \times 120 = 100\,000m \times 200$$

$$\bar{x} = 204.2 \text{ mm}$$

The centre of mass of the remaining plate is 204.2 mm from PS and 127.4 mm from PQ and at the mid-thickness of the plate.

Example 3.6

Determine the position of the centroid of the section shown in Fig. 3.18.

Solution

The section will be considered as consisting of a rectangle plus a triangle and minus the square cut-out.

Let the centroid of the section be at G, distance $\bar{x}$ from YY and $\bar{y}$ from XX.

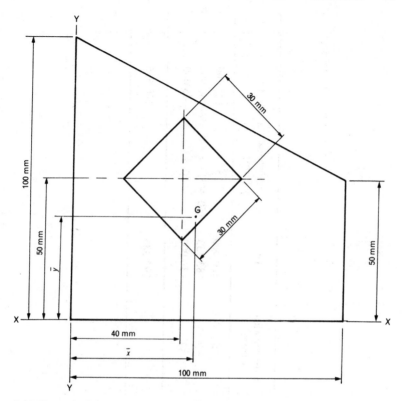

3.18 Example 3.6

Then by moments about axis YY:

rectangle   triangle   square
$(50 \times 100) \times 50 + (\tfrac{1}{2} \times 50 \times 100) \times \tfrac{100}{3} - (30 \times 30) \times 40$

$$= (50 \times 100 + \tfrac{1}{2} \times 50 \times 100 - 30 \times 30)\bar{x}$$

$$250\,000 + 83\,333 - 36\,000 = (5000 + 2500 - 900)\bar{x}$$

$$\bar{x} = \frac{297\,333}{6600} = 45 \text{ mm}$$

In the above equation the first term (in brackets) is the area of the section and the second term is the moment arm.

SECTION	AREA A [mm²]	MOMENT ARM x [mm]	Ax	MOMENT ARM y [mm]	Ay
Rectangle	$50 \times 100 = 5000$	50	250 000	25	125 000
Triangle	$\frac{1}{2} \times 50 \times 100 = 2500$	$\frac{100}{3}$	83 333	$(50 + \frac{50}{3}) = \frac{200}{3}$	166 666
Square	$-(30 \times 30) = -900$	40	−36 000	50	−45 000
Sum Σ	6600		297 333		246 666

Similarly by moments about axis XX:

rectangle   triangle   square
$(50 \times 100) \times 25 + (\tfrac{1}{2} \times 50 \times 100)(50 + \tfrac{50}{3}) - (30 \times 30) \times 50$

$$= (5000 + 2500 - 900)\bar{y}$$
$$\bar{y} = 37.4 \text{ mm}$$

The position of the centroid is 45 mm from YY and 37.4 mm from XX. This means that the centroid lies within the cut-out square section.

Calculations to determine the position of the centre of mass or centroid are frequently simplified by using a tabular form. This is now carried out for the above problem so that a comparison between the solutions can be made.

Then $\bar{x} = \dfrac{\sum Ax}{\sum A} = \dfrac{297\,333}{6600} = 45 \text{ mm}$

$\bar{y} = \dfrac{\sum Ay}{\sum A} = \dfrac{246\,666}{6600} = 37.4 \text{ mm}$

The negative sign for the area of the square arises because this section is removed.

The use of the tabular method is strongly recommended because of its simple layout.

Example 3.7

Determine the position of the centre of mass of the solid shown in Fig. 3.19 relative to the given axes.

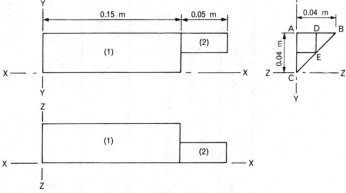

3.19 Example 3.7

PART	VOLUME V [m³]	x [m]	V_x	y [m]	V_y	z [m]	V_z
1	$\left(\frac{1}{2} \times 0.04 \times 0.04 \times 0.15\right) = \frac{120}{10^6}$	$\frac{1}{2} \times 0.15$	$\frac{9}{10^6}$	$\frac{2}{3} \times 0.04$	$\frac{3.2}{10^6}$	$\frac{1}{3} \times 0.04$	$\frac{1.6}{10^6}$
2	$(0.02 \times 0.02 \times 0.05) = \frac{20}{10^6}$	$0.15 + \frac{0.05}{2}$ $= 0.175$	$\frac{3.5}{10^6}$	$0.02 + \frac{0.02}{2}$ $= 0{\cdot}03$	$\frac{0.6}{10^6}$	$\frac{0.02}{2} = 0.01$	$\frac{0.2}{10^6}$
Σ	$\frac{140}{10^6}$	—	$\frac{12.5}{10^6}$	—	$\frac{3.8}{10^6}$	—	$\frac{1.8}{10^6}$

Solution

The gravitational force on any section of the solid is proportional to its mass. As the density of the material is constant, the mass will be proportional to the volume. Therefore, we need only deal with volumes when calculating the centre of mass position.

The solid comprises a triangular prism (1) and a square prism (2).

$$\text{Length of side of square prism} = AD$$
$$= \tfrac{1}{2} AB$$
$$= 0.02 \text{ m}$$

In the following table x, y, and z ordinates are the distances of the centre of mass of the solid considered along the XX, YY, and ZZ axis respectively.

Then
$$\bar{x} = \frac{\sum Vx}{\sum V}$$
$$= \frac{12.5}{140} = 0.0893 \text{m} = 89.3 \text{ mm}$$

$$\bar{y} = \frac{\sum Vy}{\sum V}$$
$$= \frac{3.8}{140} = 0.0271 \text{ m} = 27.1 \text{ mm}$$

$$\bar{z} = \frac{\sum Vz}{\sum V}$$
$$= \frac{1.8}{140} = 0.0128 \text{ m} = 12.8 \text{ mm}$$

The centre of mass of the solid has co-ordinates of 89.3 mm, 27.1 mm, and 12.8 mm relative to the XX, YY, and ZZ axis respectively.

Problems

1. A uniform beam XY is 5 m long and is simply supported at its ends. The beam carries loads of 400 N at a position 0.8 m from X, 250 N at a position 4 m from X, and 160 N at a position 0.6 m from Y. The value of the support reaction at X is found to be 730 N. Calculate the mass of the beam and the magnitude of the support reaction at Y.

2. A horizontal beam AB has a uniform section and is 12 m long. It rests on two supports at C and D, 8 m apart. The support at C is 2 m from end A. When the beam carries loads of 500 N at A, 1000 N at B, and 1000 N at a position 4.5 m from A the support reaction at D is found to be 1925 N. Determine:

(a) the mass of the beam;
(b) the additional load that must be applied at a distance of 4.5 m from A to produce equal support reactions at C and D.

3. A uniform beam PQ is 12 m long and its mass per metre length is 14 kg. The beam is supported at R and S which are 2 m and 1 m from P and Q respectively and carries loads of 1020 N at P, 500 N at a position 8 m from P, and 600 N at Q. Determine:

(a) the magnitude of the support reactions at R and S;
(b) the position at which a load of 820 N should be applied in order to make the support reactions equal.

4. A horizontal beam AB, of uniform section, is 8 m long and has a mass of 50 kg/m length. It rests on two supports, one at end A and the second at a point C, 6.5 m from the end A. Three vertical point loads are carried on the beam as follows: 800 N, 3 m from A; 14 000 N, 5.5 m from A; 4000 N at B. Calculate:

(a) the reaction at each support;
(b) the maximum load that can be applied at B without loss of equilibrium, assuming that the support at A cannot provide a downward reaction.

5. A uniform beam AB is 3 m long and is hinged at A to a vertical wall so that the beam is free to turn in a vertical plane. The beam has a total mass of 85 kg and a load of 550 N is applied at the free end B. The beam is maintained in a horizontal position by a chain attached to the beam at a position 2.4 m from A and to a point on the wall which is 3.2 m vertically above the hinge. Determine the tension in the chain and the magnitude and direction of the hinge reaction.

6. Fig. 3.20 shows two rigid links AB and BC which are hinged to supports at A and C and pinned together at B. Determine the vertical and horizontal components of the reactions at A and C for the loading shown.

7. A beam AB has a uniform section and is 8 m long. It is hinged to a wall at A and maintained in a horizontal position by means of a chain

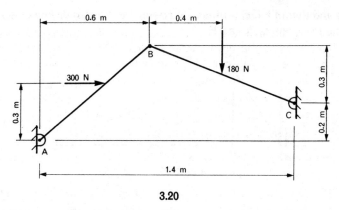

3.20

CD attached to the beam at point C, 6 m from A, such that the angle ACD is 45°. When the beam carries a load of 2 kN at B and 1.4 kN at a position 2 m from A the tension in the chain is found to be 7.2 kN. Determine the mass of the beam and the magnitude and direction of the reaction at the hinge A.

8. Two rigid links AB and BC are hinged to supports at A and C and are connected by a pin joint at B, as shown in Fig. 3.21. The lengths of AB and BC are 3 m and 5 m respectively. When a load of 150 N is applied at the mid-length position of AB and a load of 240 N is applied at a position one-third of the length of BC measured from B, calculate the magnitude of the reactions at A, B, and C. Ignore the mass of the links.

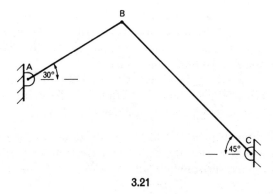

3.21

9. Two triangular plates ABC and CDE, of uniform thickness, are hinged at A, C, and E as shown in Fig. 3.22. The mass of plate ABC is

8 kg and that of CDE is 14 kg. Determine the magnitude and direction of the reactions at A and E.

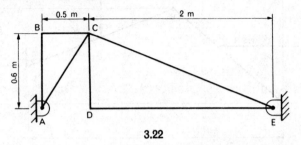

3.22

10. An electric motor is used to drive a shaft by means of a belt. If the effective diameter of the motor pulley is 80 mm and the tensions on the tight and slack sides of the belt are 725 N and 60 N respectively, what is the output torque of the motor?

11. A torque is applied to a shaft by means of an arm whose length is 60 mm. The shaft is turned against the action of a spring whose resistance is 0.15 N m per degree of rotation. What force must be applied to the arm to turn it through an angle of 90°?

12. The tensions on the tight and slack sides of a belt drive to a pulley of diameter 0.24 m are 700 N and 60 N respectively. Calculate the torque transmitted to the pulley.

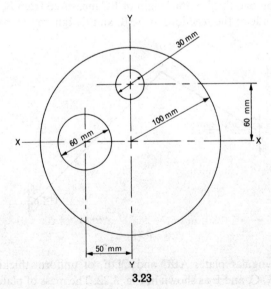

3.23

13. Determine the position of the centre of mass of the uniform plate shown in Fig. 3.23 relative to axes XX and YY.

14. A metal plate of uniform thickness is of rectangular shape with sides of 0.18 m and 0.10 m respectively. It is required to cut a hole of 0.06 m diameter in the plate such that the position of the centre of mass moves 2.5 mm towards the 0.18 m side and 5 mm towards the 0.10 m side. At what position should the centre of the cut out be marked?

15. Determine the position of the centroid of the section shown in Fig. 3.24 relative to (*a*) AB and (*b*) AC.

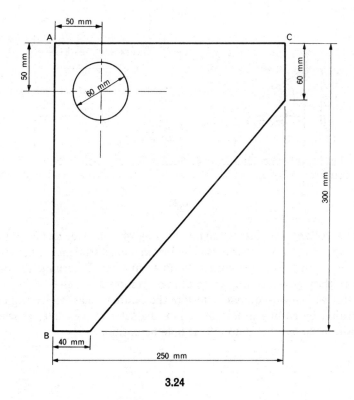

3.24

16. Determine the position of the centre of mass of the solid shown in Fig. 3.25 relative to XX, YY, and ZZ.

17. Determine the position of the centre of mass of the solid shown in Fig. 3.26 with reference to axes XX, YY, and ZZ.

60 Mechanical engineering science

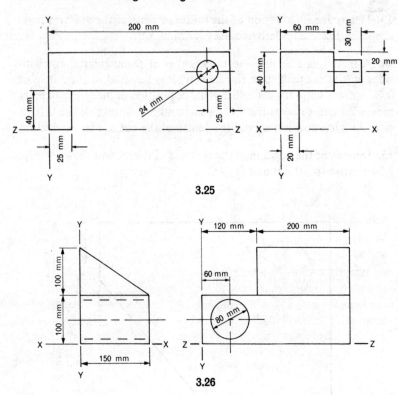

3.25

3.26

18. A rectangular plate is 350 mm long by 240 mm wide and is of uniform thickness. A corner is removed by making a straight cut between points 75 mm from the corner on adjacent sides. Determine the distance through which the centre of mass is moved.

If it were now required to restore the centre of mass to its original position by cutting a hole of 70 mm diameter in the plate, at what position should the centre of the hole be positioned?

4

Stress and strain

4.1 Transmission of forces

When a structure or mechanism transmits loads, or forces, from one point to another, the load path can often be determined by analysis or by experiment. The determination of the load path is necessary so that engineering designers can avoid high concentration of loading within the material which would otherwise lead to the onset of cracks or excessive distortion. In order to consider this problem of internal loading and distortion it is necessary to define two quantities: (*a*) stress, (*b*) strain. These are important when materials are loaded in the elastic or plastic ranges.

4.2 Direct stress

A material which is subjected to an applied load is said to be in a state of stress. Consider the simple case of a tie bar carrying an axial load, as shown in Fig. 4.1.

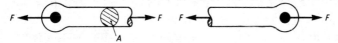

4.1 Direct stress—forces on a tie bar

In the tie bar, the force F is transmitted from one end to the other. This means that every cross section throughout the length of the bar carries the force F. It is obvious that a tie bar of large cross-sectional area will be capable of carrying a larger force than a smaller bar of

62 Mechanical engineering science

similar material and the stress is a measure of the intensity of the transmitted load.

$$\text{Direct stress} = \frac{\text{applied load}}{\text{cross-sectional area}}$$

$$\sigma = \frac{F}{A}$$

The units of stress are N/m^2; kN/m^2; etc.

This type of stress is called direct or normal stress because the plane of the section transmitting the load is normal to the axis of the load. Thus, in any given bar, an increase in the load or a reduction in the cross-sectional area will cause an increase of stress within the material. If the applied load tends to increase the length of the bar the stress is referred to as tensile, while if the applied load tends to decrease the length of the bar the stress is said to be compressive.

4.3 Strain

When a load is applied to a solid material a deformation is produced. This deformation may not be apparent to the human eye but it is always present and can be detected with a suitable measuring instrument. The strain of the body is a measure of the deformation.

$$\text{Strain} = \frac{\text{change in length}}{\text{original length}}$$

$$\varepsilon = \frac{\delta}{l}$$

As strain is the ratio of two lengths it has no units.

Thus, a strain of 0.01 means that a bar 1 m long will extend by 0.01 m, a bar 2 m long will extend by 0.02 m, and a bar 0.015 m long will extend by $0.015 \times 0.01 = 0.000\ 15$ m.

If a material returns to its original length after being strained then the imposed strain is said to be elastic, i.e. the elastic limit (see Section 4.6) has not been exceeded. However, referring to Fig. 4.3 if the strain exceeds that at which the elastic limit is reached then a permanent set will remain upon removal of the load, i.e. the material does not return to its original length. This is because the material has entered the plastic range where the total strain can no longer be recovered.

4.4 Hooke's law

This states that providing a material is worked within its elastic limit, the strain produced is directly proportional to the stress producing it.

4.5 Modulus of elasticity

If a material is loaded such that Hooke's law is satisfied, i.e. the elastic limit is not exceeded, then

$$\text{Direct stress} \propto \text{strain}$$

or

$$\frac{\text{Direct stress}}{\text{Strain}} = \text{constant}$$

This constant is known as the 'Modulus of Elasticity' or 'Young's Modulus' and is denoted symbolically by E. Thus,

$$E = \frac{\sigma}{\varepsilon}$$

and it will be seen that the units of E will be identical to the units of stress.
Typical values of E are given in Table 1 (page 67).

Example 4.1

In a tensile test on a steel bar of 12 mm diameter an axial force of 28 kN produced an extension of 0.062 mm over a gauge length of 50 mm, the load being within the elastic limit of the material.
Calculate the stress, strain, and modulus of elasticity of the material.

Solution

$$\text{Stress } \sigma = \frac{F}{A} = \frac{28 \times 10^3}{\pi/4 \times 12^2} \frac{[\text{N}]}{[\text{mm}^2]}$$

$$= 248 \text{ N/mm}^2$$
$$= 248 \times 10^6 \text{ N/m}^2$$
$$= 248 \text{ MN/m}^2$$

$$\text{Strain } \varepsilon = \frac{\delta}{l} = \frac{0.062}{50}$$

$$= 0.001\,24$$

Modulus of elasticity $E = \dfrac{\sigma}{\varepsilon}$

$$= \dfrac{248}{0.001\ 24}\ \text{MN/m}^2$$
$$= 200 \times 10^3\ \text{MN/m}^2$$
$$= 200\ \text{GN/m}^2$$

The stress is 248 MN/m², the strain 0.001 24, and the modulus of elasticity 200 GN/m².

Example 4.2

A steel bar AC is 0.3 m long and has a diameter of 20 mm for a length AB of 0.2 m and a diameter of 30 mm for the remaining length BC. Calculate the stress in the two sections when a pull of 26 kN is applied to the bar and also determine the total extension of the bar under this loading.

$$E = 210\ \text{GN/m}^2$$

Solution

A diagram is given in Fig. 4.2.

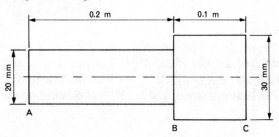

4.2 Example 4.2

$$\text{Stress in section AB} = \dfrac{26\ 000}{\pi/4 \times 20^2}\ \dfrac{[\text{N}]}{[\text{mm}^2]}$$
$$= 82.7\ \text{N/mm}^2$$
$$= 82.7\ \text{MN/m}^2$$
$$\text{Stress in section BC} = \dfrac{26\ 000}{\pi/4 \times 30^2}\ \dfrac{[\text{N}]}{[\text{mm}^2]}$$
$$= 36.8\ \text{MN/m}^2$$

Extension of AB $= \dfrac{\sigma l}{E} = \dfrac{82.7 \times 10^6 \text{ [N/m}^2\text{]} \times 0.2 \text{ [m]}}{210 \times 10^9 \text{ [N/m}^2\text{]}}$

$= 0.0787 \times 10^{-3}$ m

$= 0.0787$ mm

Extension of BC $= \dfrac{36.8 \times 10^6 \times 0.1}{210 \times 10^9}$ m

$= 0.0175$ mm

∴ Total extension of AB $= 0.0787 + 0.0175$

$= 0.0962$ mm

The stresses in AB and BC are 82.7 MN/m² and 36.8 MN/m² respectively and the total extension of AB is 0.0962 mm.

4.6 Stress–strain graph or load–extension graph

If a tensile test to destruction is carried out on a material specimen, and measurements of load and extension are taken at regular intervals, a graph of load against extension or stress against strain can be plotted. Mild steel has a characteristic stress–strain graph of the form shown in Fig. 4.3.

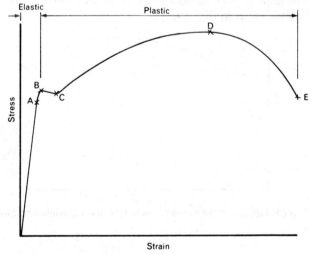

4.3 Typical stress–strain graph for mild steel

Point A indicates the limit of proportionality of the material, i.e. the point up to which the stress is proportional to the strain. For a steel, the limit of proportionality is very close to the elastic limit, point B.

Providing that point B is not exceeded the material will return to its original length upon removal of the load, i.e. the strain is elastic up to this stage, but if point B is exceeded then the material will not return to its original length and a permanent set will remain upon removal of the load. The range beyond point B is known as the plastic range.

After point B is reached an extension occurs without any increase in load (in fact, there may even be a small decrease in load), and the material is said to yield. This continues to C, at which further load must be applied to increase the extension.

Although yielding does not occur in all ductile materials it is a characteristic of irons and low carbon and alloy steels, and it is worth noting that, in certain circumstances, with such materials the extension over the range BC can be considerably greater than that produced up to B.

From C, a general curve is obtained until the maximum load is reached at D and up to this stage no marked geometrical change is noticeable in the specimen. However, just beyond D the specimen waists rapidly at one position along its length and the load required to maintain the extension also decreases until fracture finally occurs at E.

In the case of non-ferrous materials and high carbon steels no definite yield point is obtained and a typical stress–strain curve for such materials is given in Fig. 4.4.

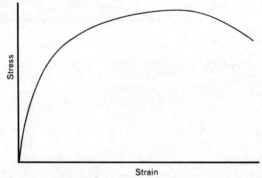

4.4 Typical stress–strain curve for non-ferrous materials and high carbon steel

4.7 Factor of safety

If a structural component is to be satisfactory in operation it is essential that the yield stress is not exceeded and a factor of safety is therefore employed in design calculations.

Stress and strain 67

$$\text{Factor of safety} = \frac{\text{yield stress}}{\text{working stress}}$$

Note: In some texts tensile strength is quoted in the above formula instead of yield stress but this has little practical significance, and it will not be used in this book.

In aircraft structures the actual load is multiplied by 1.5 or 1.33, depending upon whether the aircraft is for civil or military purposes, to give a 'factored load'. The working stresses are then based upon the factored loads.

4.8 Tensile strength

This is the stress obtained by dividing the maximum load applied during a tensile or compressive test by the original cross-sectional area of the material.

$$\text{Tensile strength} = \frac{\text{maximum load}}{\text{original cross-sectional area}}$$

Typical values for the tensile strength of materials are given in Table 1.

Table 1. **Typical values for the tensile strength and modulus of elasticity of materials**

MATERIAL	TENSILE STRENGTH MN/m^2	MODULUS OF ELASTICITY GN/m^2
Mild steel	450	205
Carbon steel	590	205
Nickel steel	1220	205
Stainless steel	1275	195
Aluminium alloy	320–480	70
Magnesium alloy	340	45
Brass	420	100
Bronze	540	120
Titanium alloy	1020	105
Oak wood	125	12.5
Polythene	15	—

As there are many types of steels, aluminium alloys, etc. the above values are quoted only as an indication of the magnitude of the property.

68 Mechanical engineering science

Example 4.3

A tube having an external diameter of 20 mm and a wall thickness of 1.5 mm is to be used as a tie bar. The yield stress of the steel is 430 MN/m² and a safety factor of 6 is required, based upon this stress.

Determine the maximum load that the tie bar can carry and the extension of a 1 m length of the tube at this loading. The modulus of elasticity is 205 GN/m².

Solution

Inside diameter of tube = $20 - 2 \times 1.5 = 17$ mm

$$\text{Area of tube} = \frac{\pi}{4}(20^2 - 17^2) = 87.2 \text{ mm}^2$$

$$= 87.2 \times 10^{-6} \text{ m}^2$$

$$\text{Maximum working stress} = \frac{\text{yield stress}}{\text{safety factor}}$$

$$= \frac{430}{6} = 71.7 \text{ MN/m}^2$$

Maximum load that can be applied to tie bar

$$= \text{working stress} \times \text{area}$$

$$= 71.7 \text{ [MN/m}^2\text{]} \times 87.2 \times 10^{-6} \text{ [m}^2\text{]}$$

$$= 6240 \times 10^{-6} \text{ MN} = 6240 \text{ N} = 6.24 \text{ kN}$$

Then

$$\varepsilon = \frac{\sigma}{E} = \frac{\delta}{l}$$

$$\delta = \frac{\sigma l}{E} = \frac{71.7 \times 10^6 \text{ [N/m}^2\text{]} \times 1 \text{ [m]}}{205 \times 10^9 \text{ [N/m}^2\text{]}}$$

$$= 0.350 \times 10^{-3} \text{ m}$$

$$= 0.350 \text{ mm}$$

The maximum load that can be applied is 6.24 kN and the subsequent extension of a 1 m length is 0.350 mm.

4.9 Tensile test results

In addition to being able to obtain the yield stress and modulus of elasticity (from the slope of the stress–strain graph up to the limit of pro-

Stress and strain   69

portionality) we can also calculate the tensile strength, percentage elongation and percentage reduction in area from the results of a tensile test to fracture.

$$\text{Tensile strength} = \frac{\text{maximum load}}{\text{original cross-sectional area}}$$

$$\text{Percentage elongation} = \frac{\text{increase in length}}{\text{original length}} \times 100$$

$$\text{Percentage reduction in area} = \frac{\text{original area} - \text{final area at fracture}}{\text{original area}} \times 100$$

The elongation and reduction in area of a material give an indication of its ductility, which is its ability to withstand strain without fracture occurring. As the percentage elongation depends upon the original gauge length used it is customary, for purposes of comparison, to use a gauge length as specified by the British Standards Institution.

Example 4.4

In a tensile test on a steel bar the following results were obtained:

Load (kN)	10	20	30	40	50	54	60
Extension (mm)	0.020	0.041	0.061	0.082	0.103	0.118	1.43

Initial diameter of bar = 15 mm
Final diameter of bar at fracture = 9.03 mm
Initial gauge length = 75 mm
Final gauge length = 97.3 mm
Maximum load applied during test = 87.3 kN

From these results determine:

(a) the modulus of elasticity;
(b) the yield stress;
(c) the tensile strength;
(d) the percentage elongation;
(e) the percentage reduction in area.

Solution

A graph of load against extension is plotted in Fig. 4.5.

70 Mechanical engineering science

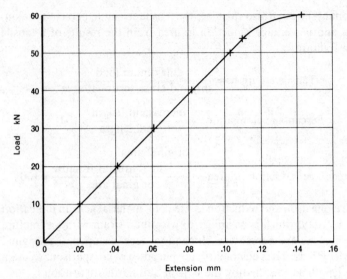

4.5 Load–extension graph—Example 4.4

(a) Modulus of elasticity,

$$E = \frac{\text{stress}}{\text{strain}}$$

$$= \frac{F}{A} \cdot \frac{l}{\delta}$$

$$= \frac{l}{A} \times \text{gradient of load–extension graph}$$

$$= \frac{75 \times 4}{\pi \times 15^2} \times \frac{48.5 \times 10^3}{0.1}$$

$$= 206 \times 10^3 \text{ N/mm}^2$$

$$= 206 \text{ GN/m}^2$$

(b) From the graph, the load at which yielding occurred is estimated to be 55 kN.

$$\therefore \text{Yield stress} = \frac{55 \times 10^3}{\frac{\pi}{4} \times 15^2} \frac{[\text{N}]}{[\text{mm}^2]}$$

$$= 311 \text{ N/mm}^2 = 311 \text{ MN/m}^2$$

(c) Tensile strength

$$= \frac{\text{maximum load}}{\text{cross-sectional area}}$$

$$= \frac{87.3 \times 10^3}{\frac{\pi}{4} \times 15^2} \frac{[\text{N}]}{[\text{mm}^2]}$$

$$= 494 \text{ N/mm}^2 = 494 \text{ MN/m}^2$$

(d) Percentage elongation

$$= \frac{\text{final length} - \text{initial length}}{\text{initial length}} \times 100$$

$$= \left(\frac{97.3 - 75}{75}\right) \times 100$$

$$= 29.7\%$$

(e) Percentage reduction in area

$$= \frac{\text{original area} - \text{final area}}{\text{original area}} \times 100$$

$$= \left(\frac{\frac{\pi}{4} \times 15^2 - \frac{\pi}{4} \times 9.03^2}{\frac{\pi}{4} \times 15^2}\right) \times 100$$

$$= \left(1 - \frac{9.03^2}{15^2}\right) \times 100$$

$$= 63.7\%$$

4.10 Shear stress

Stresses other than direct stress can arise in a material when it is loaded. A frequent case is that of a material which is loaded in such a way that adjacent layers of the material tend to slide relative to each other. Such an action is known as shearing and examples are illustrated in Fig. 4.6. In each of the two cases shown it will be seen that the material on either side of XX moves, or has a tendency to move, in opposite directions. Thus, the material is in shear and although the shear stress varies over the area under shear we can obtain the average shear stress from

$$\text{Average shear stress} = \frac{\text{shearing force}}{\text{area resisting shear}}$$

72 Mechanical engineering science

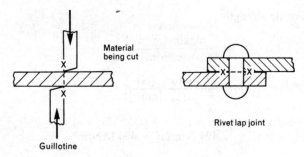

4.6 Examples of shearing action

Example 4.5

The fork joint shown in Fig. 4.7 is required to transmit a pull of 42 kN. If the diameter of the pin is 15 mm determine the shear stress in the pin.

What is the safety factor if the pin is made from steel having a yield shear stress of 190 MN/m²?

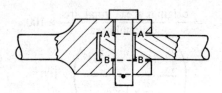

4.7 Fork joint—Example 4.5

Solution

From the diagram it will be seen that the pin is in shear across both AA and BB, i.e. it is in double shear.

$$\therefore \text{Area resisting shear} = 2\left(\frac{\pi \times 15^2}{4}\right) = 354 \text{ mm}^2$$

$$\text{Average shear stress in pin} = \frac{\text{shearing force}}{\text{area resisting shear}}$$

$$= \frac{42 \times 10^3}{354} \frac{[\text{N}]}{[\text{mm}^2]}$$

$$= 118.8 \text{ N/mm}^2$$

$$= 118.8 \text{ MN/m}^2$$

Stress and strain 73

$$\text{Safety factor} = \frac{\text{shear yield stress}}{\text{working shear stress}}$$

$$= \frac{190}{118.8}$$

$$= 1.6$$

The shear stress in the pin is 118.8 MN/m² and the safety factor at this stress is 1.6.

Example 4.6
A duralumin strut has an internal diameter of 30 mm and a length of 1.1 m. The strut forms part of a framework and the maximum axial load to which it is subjected is 80 kN. At this loading, a safety factor of 3 is specified, based upon a compressive yield stress of 255 MN/m². Calculate the minimum allowable thickness of the strut wall.

Solution

$$\text{Allowable working stress} = \frac{\text{compressive yield stress}}{\text{safety factor}}$$

$$= \tfrac{255}{3} = 85 \text{ MN/m}^2$$

Let the minimum wall thickness required be t [mm]. Then, the external diameter of the strut is $(30+2t)$ mm. Now,

$$\text{Stress} = \frac{\text{applied force}}{\text{area}}$$

$$\therefore \text{Area required} = \frac{80 \times 10^3}{85 \times 10^6} \frac{[\text{N}]}{[\text{N/m}^2]}$$

$$= 0.942 \times 10^{-3} \text{ m}^2$$

$$= 942 \text{ mm}^2$$

$$\therefore 942 = \frac{\pi}{4}[(30+2t)^2 - 30^2]$$

$$(30+2t)^2 - 30^2 = 942 \times \frac{4}{\pi}$$

$$= 1200$$

74   Mechanical engineering science

$$(30+2t)^2 = 1200+900$$
$$(30+2t) = 45.8$$
$$\therefore t = 7.9 \text{ mm}$$

The minimum thickness of the tube wall is 7.9 mm.

Problems

1. What is meant by the terms: 'Hooke's Law' and 'Modulus of Elasticity'?

A tie rod 6 m long and 1200 mm² in cross-section, increases in length by 2.9 mm when carrying a tensile load of 115 kN. Calculate:
(a) the tensile stress in the rod;
(b) the strain;
(c) the modulus of elasticity;
(d) the factor of safety, if the tensile yield stress of the material is 320 MN/m².

2. A hollow steel tube having an outside diameter of 30 mm is to be used as a strut. The strut is required to withstand a load of 18.5 kN. If there is to be a safety factor of 6 what is the maximum permissible internal diameter of the tube?

What shortening of the strut would occur on a 3.6 m length under this loading? (Compressive yield stress = 300 MN/m²; $E = 210$ GN/m².)

3. A mild steel column has an outer diameter of 0.22 m and is 3 m long. When subjected to an axial compressive load of 895 kN there is a shortening of 2 mm. Calculate the wall thickness of the column. ($E = 203$ GN/m².)

4. (a) Define: (i) stress, (ii) strain, (iii) limit of proportionality.

(b) Two specimens of apparently the same material were tested in tension. The first, of diameter 15 mm, extended 0.36 mm on a gauge length of 0.18 m under a load of 36 kN. The second, of diameter 25 mm, extended 0.25 mm on a gauge length of 0.24 m under a load of 52 kN. Do these tests support the assumption that the specimens are of the same material?

5. A brass bar ABC is 0.24 m long and has two sections. Section AB of 20 mm diameter and 0.14 m long and section BC which is 25 mm

square. When a tensile loading of 30 kN is applied to the bar what will be the stresses in the two sections? Calculate also the total extension of the bar. (Modulus of elasticity of brass is 112 GN/m².)

6. A steel bar of square section has a total length of 0.7 m. It has a side of 30 mm for a portion of its length and one of 17.5 mm for the remainder. When subjected to an axial load the extension of each portion is 0.12 mm. If the modulus of elasticity for the material is 205 GN/m² calculate:

(a) the length of each section of the bar;
(b) the stress in each section of the bar.

7. A tensile test to fracture carried out on a steel test specimen gave the following results.

Diameter of test bar = 10 mm
Length between gauge points = 50 mm

Load (kN)	2	6	12	18
Extension (mm)	0.005	0.017	0.037	0.054
Load (kN)	24	26	27	34
Extension (mm)	0.072	0.078	0.96	4.2

Maximum load applied during test = 43.6 kN
Final distance between gauge points = 65.5 mm
Diameter of bar at fracture = 7.62 mm

From these results determine:

(a) the modulus of elasticity for the material;
(b) the stress at the elastic limit;
(c) the tensile strength;
(d) the percentage elongation;
(e) the percentage reduction in area.

8. State 'Hooke's law'. A tensile test on a structural steel bar showed a yielding load of 70.6 kN on a specimen 14 mm diameter. Calculate the extension which would occur, under working load, on a 3 m-long tie rod of this material if a design factor of safety of 5 is used and the modulus of elasticity for the material is 206 GN/m².

9. A hydraulic cylinder has an internal diameter of 0.22 m and the cylinder cover is to be held in position by six bolts. If the tensile yield strength of the bolt material is 360 MN/m², determine the minimum core diameter of the bolts required to withstand a maximum cylinder pressure of 4100 kN/m² if there is to be a factor of safety of 4.

76 Mechanical engineering science

10. A brass journal bearing has an internal diameter of 30 mm and an external diameter of 55 mm. If the length of the bearing is 0.14 m and the maximum shortening that can be allowed is 0.05 mm, what is the greatest axial thrust that can be applied? Calculate also, the stress and safety factor at this loading. ($E = 110$ GN/m^2; tensile yield stress $= 470$ MN/m^2.)

11. (a) Define (i) *modulus of elasticity* and (ii) *yield point*.
(b) A mild steel specimen of diameter 16 mm was found to have an extension of 0.138 mm on a 0.2 m gauge length when a load of 28 kN was applied. Assuming the load to lie within the limit of proportionality determine the modulus of elasticity of the material. If a load of 60 kN produced an extension of 0.384 mm on a 0.25 m gauge length, determine whether the loading would still be within the limit of proportionality.

12. The following results were obtained during a tensile test to destruction on a steel test specimen having a diameter of 15.95 mm and a gauge length of 0.08 m.

Extension for load of 40 kN $= 0.076$ mm
Maximum load applied during test to fracture $= 93$ kN
Diameter of test specimen at fracture $= 13.04$ mm
Final length between gauge points $= 0.1062$ m

Determine from these results:

(a) the modulus of elasticity;
(b) the tensile strength;
(c) the percentage reduction in area;
(d) the percentage elongation.

13. A tensile test to destruction carried out on a hard drawn brass test bar provided the following results:

Load (kN)	10	20	24	28	31.5
Extension (mm)	0.047	0.093	0.112	0.13	0.15
Load (kN)	34	36	38	40	
Extension (mm)	0.168	0.195	0.23	0.49	

Maximum load applied during test $= 55.3$ kN
Initial diameter of test bar $= 14$ mm
Final diameter of bar at fracture $= 7.97$ mm
Initial gauge length $= 75$ mm
Final gauge length $= 122.2$ mm

Plot a graph of the results and determine:
(a) the stress at the limit of proportionality;
(b) the modulus of elasticity;
(c) the tensile strength;
(d) the percentage elongation;
(e) the percentage reduction in area.

14. Two duralumin plates are held together by three 7.5 mm diameter rivets of the same material. If the yielding shear stress of duralumin is 180 MN/m^2 and the rivets are to have a safety factor of 4 what is the maximum load that can be carried by the joint?

15. An engine piston is 60 mm diameter and at the top dead centre position is subjected to a maximum cylinder pressure of 2 MN/m^2. The gudgeon pin is 20 mm outside diameter and 14 mm inside diameter and the minimum cross-sectional area of the connecting rod is 360 mm^2.

Determine the shear stress in the gudgeon pin and the maximum compressive stress in the connecting rod.

16. It is required to punch holes of 45 mm diameter in a mild steel plate of thickness 3 mm. If the shear yield stress of the steel is 250 MN/m^2 calculate the punching force required.

17. Two tubes are fitted with fork ends and are joined together by a steel pin having a diameter of 20 mm. Calculate the shear stress in the pin when the joint is transmitting a load of 42.5 kN, and the factor of safety if the shear yield stress is 1.2 GN/m^2.

18. A steel bolt having a 0 B.A. thread passes through a rigid material of thickness of 0.1 m and the nut is hand tightened so that no slackness is present. Calculate the angle through which the nut must now be tightened so that a stress of 840 MN/m^2 is induced in the bolt. (Modulus of elasticity = 204 GN/m^2; a 0 B.A. thread has a diameter of 6 mm and a pitch of 1 mm.)

5

Friction

5.1 Friction

Whenever two surfaces move, or tend to move, relative to one another a force is brought into action which tends to oppose the motion between the surfaces. This opposing force is the result of friction between the surfaces.

Consider a body resting on a horizontal surface. If the surfaces were perfectly smooth then only forces normal to the contact surfaces could be present and no force would be required to move the body horizontally (Fig. 5.1(i)). In practice, however, no such state of affairs exists and, if a gradually increasing lateral force P is applied to the body, motion will not occur until the force P is greater than the limiting frictional resisting force F (Fig. 5.1(ii)).

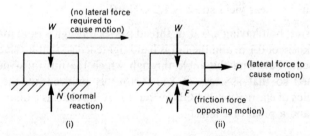

5.1 (i) Smooth surface   (ii) Rough surface

Let us now examine this state of affairs more closely. When the magnitude of P is insufficient to move the body the frictional resisting force F is exactly equal to P. Thus, the reaction R from the surface provides a vertical component N, to balance the gravitational force on the body, W, and a horizontal component F, which balances P. This is shown in

Fig. 5.2(i). As P is gradually increased so the angle θ gradually increases. There is, however, a limit to the horizontal reaction F which can be created, and when this arises F is said to have reached its limiting value. A further increase in P will cause motion between the surfaces.

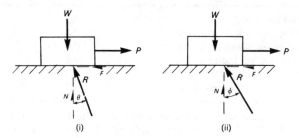

5.2 (i) Block stationary—
P gradually increasing

(ii) Body on point of moving—limiting friction

When the body is on the point of slipping:

for horizontal equilibrium: $P = F = R \sin \phi$
for vertical equilibrium: $W = N = R \cos \phi$

The angle ϕ, Fig. 5.2(ii), is known as the angle of limiting friction.

It has been found experimentally that once motion has started the magnitude of the frictional resisting force is slightly reduced. However, as the difference is small it will not be dealt with further at this stage.

5.2 Coefficient of friction

From experimental work on dry surfaces it has been found that there is a relationship between the frictional force F and the normal reaction N between the surfaces, i.e.

$$F \propto N$$

or

$$\frac{F}{N} = \text{constant} = \text{coefficient of friction } \mu$$

The value of μ depends more upon the character of the surfaces in contact rather than the actual materials, but it is not likely to exceed 1.

An average value can be obtained experimentally.

Table 2 gives some typical values for the coefficient of friction, between dry surfaces.

80 Mechanical engineering science

Table 2. **Typical values for the coefficient of friction between dry surfaces**

MATERIALS	COEFFICIENT OF FRICTION
Metal on metal	0.2
Rubber on wood or metal	0.4
Leather on wood or metal	0.4
Hardwood on metal	0.6
Asbestos on cast iron	0.45
Rubber on a road surface	0.9

The cleanliness and roughness of the surfaces involved has a considerable effect upon the coefficient of friction and values differing widely from those quoted above can be easily obtained.

5.3 Laws of friction

The following statements have been found to be generally true as a result of experimental work on dry surfaces.

(a) The frictional force always opposes motion, is proportional to the normal force between the surfaces, and is very dependent on the character and surface finish of the materials involved.

(b) The frictional force is independent of the area of contact between the surfaces.

(c) Providing that the sliding speed between the surfaces is relatively low, the frictional force is found to be independent of the sliding speed.

These are not fundamental 'laws' but more of a general guide. It must be emphasised that they are not necessarily valid for high pressures between the surfaces or high sliding speeds and certainly not valid for lubricated surfaces.

Example 5.1

A case of mass 300 kg is drawn along a level floor at a steady speed by means of a rope. If the coefficient of friction between the case and the floor is 0.35 calculate the pulling force required when:

(a) the rope is parallel to the floor;
(b) the rope is inclined at 15° to the horizontal floor.

Solution

(a) A diagram is given in Fig. 5.3.

Friction   81

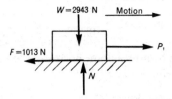

5.3 Case pulled along a level floor—Example 5.1

$$\text{Gravitational force acting on case} = 300 \times 9.81$$
$$= 2943 \text{ N}$$

Let the pulling force required be P_1.
For vertical equilibrium, the normal reaction between the surfaces $N = 2943$ newtons. Then,

$$\text{Frictional resistance } F = \mu N$$
$$= 0.35 \times 2943$$
$$= 1031 \text{ N}$$

As the case is being pulled at a steady speed there is no resultant force on it. Thus,

$$P_1 = F$$
$$= 1031 \text{ N}$$

The horizontal pulling force required is 1031 N.

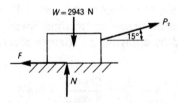

5.4 Pull inclined to horizontal—Example 5.1

(b) A diagram for this case is given in Fig. 5.4. Let the pulling force required be P_2. Resolving vertically for equilibrium:

$$N + P_2 \sin 15° = 2943$$
$$N = 2943 - P_2 \sin 15°$$

82 Mechanical engineering science

Then $\qquad F = \mu N$

$$= 0.35(2943 - P_2 \sin 15°)$$

Again, there can be no resultant horizontal force acting on the case. Therefore, resolving horizontally:

$$P_2 \cos 15° = F$$

$$= 0.35(2943 - P_2 \sin 15°)$$

$$0.966 P_2 = 1031 - 0.09 P_2$$

$$1.056 P_2 = 1031$$

$$P_2 = 977 \text{ N}$$

The pulling force required is 977 N.

It will be noticed that when the pulling force is inclined to the horizontal the force required (977 N) to cause motion is less than when the pulling force is applied horizontally (1031 N). This result is not what one might have expected. The explanation is that the vertical component of P_2 has the effect of reducing the normal reaction between the surfaces. This, in turn, means that a smaller frictional resisting force is created so requiring a lower horizontal component of the pulling force, P_2, to cause motion.

5.4 Angle of friction

When a state of limiting friction is reached the resultant reaction R was shown to act at angle ϕ to the normal reaction. By making use of this, problems can often be simplified to 'three force problems' and can therefore be solved by applying the triangle of forces. From Sections 5.1 and 5.2:

$$\mu = \frac{F}{N} = \frac{R \sin \phi}{R \cos \phi} = \tan \phi$$

ϕ is known as the angle of friction.

5.5 Inclined plane—angle of repose

Consider a body resting on a rough plane inclined at angle α to the horizontal (Fig. 5.5). In this case the body tends to move down the

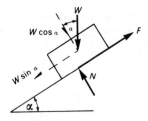

5.5 Body on an inclined plane—angle of repose

plane and therefore the frictional resisting force F acts up the plane, so as to oppose any motion. The gravitational force acting on the body, W, can be resolved into $W \cos \alpha$ perpendicular to the plane and $W \sin \alpha$ parallel to the plane.

If the angle of inclination of the plane is gradually increased the block will eventually slide down the plane.

Then, resolving perpendicular and parallel to the plane:

$$N = W \cos \alpha$$

and

$$F = W \sin \alpha$$

But

$$\mu = \frac{F}{N} = \frac{W \sin \alpha}{W \cos \alpha} = \tan \alpha$$

But from Section 5.4,

$$\mu = \tan \phi$$

$$\therefore \alpha = \phi$$

This particular value for the inclination of the plane is known as the 'angle of repose'. If $\alpha > \phi$ the body will slide down the plane but if $\alpha < \phi$ the body will remain at rest unless an external force is applied to cause motion. This fact is often employed when determining the value of μ experimentally.

Problems involving additional forces can be solved both graphically and analytically, as illustrated by Examples 5.2 and 5.3.

Example 5.2

Derive an expression for the force required to pull a body of mass m up a plane inclined at θ to the horizontal, the pulling force being inclined at angle α to the plane:

Hence obtain the force required when:

(a) the force is applied parallel to and up the plane;
(b) the force is applied horizontally such that it assists motion up the plane.

Solution

ANALYTICALLY

Consider a body of mass m to be pulled up a plane of inclination θ by a

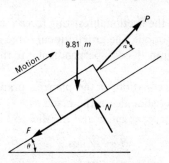

5.6 Body pulled up an inclined plane—Example 5.2

force P acting at angle α to the plane, as shown in Fig. 5.6, and let the normal reaction between the plane and the body be N.

Gravitational force on body $= 9.81m$ newtons

Resolving perpendicular to the plane:

$$N + P \sin \alpha = 9.81m \cos \theta$$

$$N = 9.81m \cos \theta - P \sin \alpha$$

The frictional force opposing motion is then given by

$$F = \mu N$$

$$= \mu(9.81m \cos \theta - P \sin \alpha) \quad (1)$$

Resolving parallel to the plane:

$$P \cos \alpha = F + 9.81m \sin \theta$$

Substituting for F from equation (1) gives

$$P \cos \alpha = \mu(9.81m \cos \theta - P \sin \alpha) + 9.81m \sin \theta$$

$$P \cos \alpha = 9.81m(\mu \cos \theta + \sin \theta) - \mu P \sin \alpha$$

Friction 85

$$P(\mu \sin \alpha + \cos \alpha) = 9.81m(\mu \cos \theta + \sin \theta)$$

$$P = \frac{9.81m(\mu \cos \theta + \sin \theta)}{(\cos \alpha + \mu \sin \alpha)} \qquad (2)$$

(a) When P is applied parallel to the plane then $\alpha = 0$. Hence, $\sin \alpha = 0$ and $\cos \alpha = 1$. Then from equation (2):

$$P = \frac{9.81m(\mu \cos \theta + \sin \theta)}{(1+0)}$$

$$= 9.81m(\mu \cos \theta + \sin \theta)$$

(b) When P is applied horizontally so that it assists motion up the plane then $\alpha = -\theta$. Thus,

$$\cos \alpha = \cos(-\theta) = \cos \theta$$

and $$\sin \alpha = \sin(-\theta) = -\sin \theta$$

Then, from equation (2):

$$P = \frac{9.81m(\mu \cos \theta + \sin \theta)}{(\cos \theta - \mu \sin \theta)}$$

GRAPHICALLY

This problem, and also Examples 5.1 and 5.3, can be solved graphically.

As the frictional force is always equal to μN, we can draw the direction of the resultant reaction R between the body and the plane at an angle ϕ to the normal to the plane, as shown in Fig. 5.7. Note that $\tan \phi = \mu$ (ref. Section 5.4).

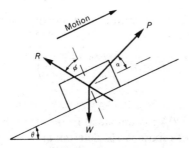

5.7 Forces acting on body—Example 5.2

As the body is in equilibrium it follows that there is no resultant force acting on it and therefore a vector triangle of forces can be drawn as shown in Fig. 5.8.

86 Mechanical engineering science

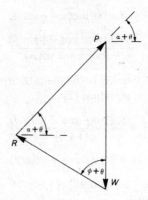

5.8 Force vector diagram—Example 5.2

By sine rule:

$$\frac{P}{\sin(\phi+\theta)} = \frac{W}{\sin\{90-(\phi+\theta)+(\alpha+\theta)\}}$$

$$\frac{P}{\sin(\phi+\theta)} = \frac{W}{\sin\{90-(\phi-\alpha)\}}$$

$$\frac{P}{\sin(\phi+\theta)} = \frac{W}{\cos(\phi-\alpha)}$$

$$P = W\left(\frac{\sin(\phi+\theta)}{\cos(\phi-\alpha)}\right)$$

$$= W\left(\frac{\sin\phi\cos\theta+\cos\phi\sin\theta}{\cos\phi\cos\alpha+\sin\phi\sin\alpha}\right)$$

Dividing both the numerator and denominator by $\cos\phi$ and putting $\tan\phi = \mu$ gives

$$P = W\left(\frac{\mu\cos\theta+\sin\theta}{\cos\alpha+\mu\sin\alpha}\right)$$

Example 5.3

A lathe of mass 1540 kg is unloaded from a lorry by sliding it down rails. If the angle of repose is 16° determine:

(a) the force required to push the lathe down the rails when they are inclined at 12° to the horizontal;

(b) the upward force that must be applied parallel to the rails to prevent the lathe from sliding down when the rails are inclined at 28° to the horizontal.

Solution

(a) A diagram for this case is given in Fig. 5.9.

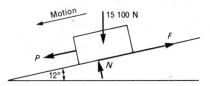

5.9 Lathe pushed down inclined rails—Example 5.3

N is the normal reaction to the rails and F is the frictional force opposing motion down the plane.

Gravitational force on lathe = $9.81 \times 1540 = 15\,100$ N

Resolving perpendicular to the rails

$$N = 15\,100 \cos 12° = 14\,780 \text{ N}$$

Then, $\qquad F = \mu N$

But $\qquad \mu = \tan 16°$

$\qquad \therefore F = \tan 16° \times 14\,780$ N

$\qquad\qquad = 4235$ N

If P is the additional force required to push the lathe down the rails then, resolving parallel to the rails:

$$P + 15\,100 \sin 12° = F$$

$$P + 3140 = 4235$$

$$P = 1095 \text{ N}$$

The required force down the plane is 1095 N.

(b) A diagram is given in Fig. 5.10 where the notation is as for (a)

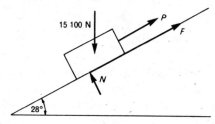

5.10 Lathe on inclined rails—Example 5.3

except that P is now the force required parallel to rails to prevent the lathe sliding down the rails.

Resolving perpendicular to the rails:

$$N = 15\,100 \cos 28° = 13\,330 \text{ N}$$

Again, $\quad F = \mu N = \tan 16° \times 13\,330$
$$= 3820 \text{ N}$$

and resolving parallel to the rails:

$$15\,100 \sin 28° = P + F$$
$$\therefore P = 3280 \text{ N}$$

The required force up the plane is 3280 N.

5.6 Friction in a journal bearing

Journal bearings are used to carry lateral loads and the shaft diameter is usually slightly less than the bearing diameter. When the shaft is stationary the contact will occur at the lowest point A (Fig. 5.11). However, when the shaft rotates it will roll up the bearing surface until limiting friction is reached at point B and slip occurs.

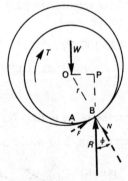

5.11 Friction in a journal bearing

It should be noted here that if the bearing is lubricated the shaft will roll in the opposite direction round the bearing, i.e. to the other side of point A. This cannot be pursued further in this text and must be left for future studies in Theory of Machines.

Returning to Fig. 5.11, the frictional resistance F acts to oppose the rotation of the shaft and N is the normal reaction at the point of contact. The resultant reaction R therefore acts at angle ϕ to the normal reaction N. Thus the forces acting on the shaft are W and R and for vertical equilibrium R must equal W and be vertically upwards. It will be seen that W and R produce a torque which opposes the rotation of the shaft.

$$\text{Friction torque} = W \times \text{OP}$$
$$= Wr \sin \phi$$

where r is the radius of the shaft.

But for small angles $\sin \phi \approx \tan \phi$. Thus,

$$\text{Friction torque} = Wr \tan \phi$$
$$= \mu Wr$$

Example 5.4

A shaft has a diameter of 0.20 m and a mass of 4200 kg. It is supported in three bearings. If the coefficient of friction between the shaft and the bearing is 0.03 find, when the shaft is rotating at 800 rev/min:

(a) the friction torque;
(b) the work done per minute in overcoming friction.

Solution

$$\text{Gravitational force on shaft} = 4200 \times 9.81$$
$$= 41\ 202 \text{ N}$$

(a) Although the shaft is supported in three bearings the problem is solved by considering all the force to act on one bearing.

From Section 5.6

$$\text{Friction torque} = \mu Wr$$
$$= 0.03 \times 41\ 202\ [\text{N}] \times 0.10\ [\text{m}]$$
$$= 123.6 \text{ N m}$$

(b) Work done by a torque $= T\theta$, where $T =$ torque and $\theta =$ angle turned through

$\therefore$ Work done against friction $= 123.6\ [\text{N m}] \times 800 \times 2\pi\ [\text{rad/min}]$
$$= 621\ 000 \text{ J/min}$$
$$= 621 \text{ kJ/min}$$

Mechanical engineering science

5.7 Advantages and disadvantages of friction

The presence of frictional resisting forces is often very desirable and in other instances is most undesirable.

We rely on friction between shoe soles and the ground to enable us to walk but let us consider a car, as an engineering product, to examine these advantages and disadvantages. The power is produced within the engine and the engineer attempts to minimise the frictional forces and so reduce the power losses. The power is then transmitted through the clutch, where high frictional forces are required between the clutch plates. The drive then passes through the gear box and rear axle where it is again essential to reduce the frictional effects to a minimum. However, a high friction force is necessary to maintain a good tyre adhesion with the road and high friction forces are again required for braking purposes.

When high frictional effects are necessary the engineer chooses the most suitable material for the particular application, and when the frictional forces are undesirable materials having a low coefficient of friction are used, e.g. nylon gears, or a lubricant is interposed between the sliding surfaces, e.g. in the engine, gear box, and rear axle of a car. Bearings are used to provide smooth running with low friction effects.

The best lubricants are oils and greases. The lubricant used depends on the pressure and the speed between the sliding surfaces. It is imperative that the lubricant used must not be squeezed out from between the bearing surfaces.

Problems

1. A loaded trolley of mass 400 kg is pulled along a level surface at a steady speed by a force of 940 N applied horizontally. What force would be required if the mass of the loaded trolley was increased to 520 kg and the pulling force then applied at 15° to the horizontal?

2. A steel casting having a mass of 2700 kg is pulled along a horizontal floor at a uniform speed by means of a force of 8750 N applied parallel to the floor. Determine the value of the coefficient of friction.

What pulling force would be required if:

(a) the inclination of the pull is at 20° to the horizontal;
(b) the casting is pulled up a ramp inclined at 18° to the horizontal with the inclination of the pulling force being at 30° to the horizontal?

3. A truck of mass 550 kg is pushed up a plane inclined at 12° to the horizontal by a force of 2800 N applied horizontally. Determine:
(a) the coefficient of friction;
(b) the force acting parallel to the plane required to draw the truck up the incline.

4. A case having a mass of 450 kg requires a force of 1000 N to pull it along a rough level surface. What additional force would be required to pull the case up an incline of 15° if the surface is similar?

5. State the factors which affect the friction between two plane dry surfaces when the relative speed between them is low.
   The table of a planing machine has a mass of 560 kg and moves horizontally on its bed. If the coefficient of friction between the table and the bed is 0.065, what force would be required to overcome friction when a casting having a mass of 900 kg is mounted on the machine table?

6. Show that the minimum force required to pull a body up a rough inclined plane is given by $[W \sin(\theta+\phi)]/\cos \phi$ where W is the gravitational force acting on the body, θ is the inclination of the plane to the horizontal and $\mu = \tan \phi$ is the coefficient of friction between the body and the plane.

7. What force, inclined at 50° to the horizontal, is required to pull a body having a mass of 12 kg up a rough plane inclined at 25° to the horizontal? Assume a coefficient of friction of 0.24.

8. A ladder having a length of 8 m and a mass of 30 kg leans against a smooth vertical wall while its lower end rests on a rough horizontal surface. When the ladder is inclined at 58° to the horizontal it is just on the point of slipping. Assuming the gravitational force on the ladder to act at its mid-length, determine the limiting coefficient of friction.

9. A body of mass 5.6 kg is placed on a rough plane which is inclined at 18° to the horizontal, the coefficient of friction between the object and the plane being 0.20. A second body of mass 4.8 kg is now placed on the plane below the first body but in contact with it. If the coefficient of friction between the second body and the plane is 0.36 show that movement will occur. Determine what the mass of the second body should be to prevent movement occurring.

10. A shaft having a diameter of 0.15 m and a mass of 860 kg is

supported in plain bearings, the coefficient of friction between the shaft and the bearings being 0.058. Determine:

(a) the friction torque;
(b) the energy expended per minute in overcoming friction when the shaft is rotating at 500 rev/min.

11. The load on a bearing supporting a shaft of 50 mm diameter is 3200 N. If the energy consumed per second in overcoming friction is 400 J when the speed of rotation of the shaft is 1400 rev/min, calculate the coefficient of friction between the shaft and the bearing.

12. A shaft and its rotor have a total mass of 900 kg and the diameter of the shaft is 0.12 m. The shaft runs in plain bearings, the coefficient of friction being 0.026. When the shaft rotates at 750 rev/min, calculate the friction torque and the work done per minute in overcoming friction.

6

Simple machines

6.1 The machine

The machine is a combination of rigid bodies for the transmission of motion, forces, and work. Its purpose is frequently to enable operations to be carried out that would otherwise be extremely difficult or even impossible to achieve manually, e.g. lifting of a large mass. In most instances a machine is designed to achieve the greatest efficiency but there are cases when the efficiency must be limited (Ref. Section 6.8).

6.2 The ideal machine

In an ideal machine no frictional resistances exist and hence the energy input to the machine will equal the energy output from the machine in the same time. In practice, however, friction is always present and so no ideal machine exists. In these circumstances, the energy input to the machine is equal to the energy output from the machine plus the energy expended in overcoming friction.

6.3 Mechanical advantage

As mentioned above, a machine is frequently used to raise a load by means of a considerably smaller effort. The mechanical advantage (M.A.) of a machine is defined by:

$$\text{M.A.} = \frac{\text{load}}{\text{effort}}$$

Thus, as the mechanical advantage is simply the ratio of two forces it has no units. The mechanical advantage, which usually has a value greater than 1, is obtained experimentally.

6.4 Velocity ratio

The velocity ratio (V.R.) of a machine is defined by:

$$\text{V.R.} = \frac{\text{distance moved by effort}}{\text{distance moved by load}}$$

The velocity ratio is the ratio of two distances and therefore, as for mechanical advantage, it has no units. However, unlike mechanical advantage, the velocity ratio depends upon the construction of a machine and is therefore constant for a given machine.

The fact that a large load can be raised, by using a machine, with a considerably smaller effort does not mean that one is 'getting more work out than was put in'. It is because the effort moves through a much greater distance than the load. Thus, the velocity ratio of a machine is frequently quite high.

6.5 Efficiency of a machine

Efficiency is defined as:

$$\text{Efficiency} = \frac{\text{work output}}{\text{work input}} \text{ (in same time)}$$

This expression can be expanded and, in the case of a machine, be written in terms of the mechanical advantage and velocity ratio. Thus,

$$\text{Efficiency} = \frac{\text{load} \times \text{distance moved by load}}{\text{effort} \times \text{distance moved by effort}}$$

$$= \frac{\text{M.A.}}{\text{V.R.}}$$

In the case of an ideal machine, where the efficiency would be 1 (or 100%), it follows that the mechanical advantage and velocity ratio would be equal, i.e.

Ideal mechanical advantage = velocity ratio

However, no machine is ideal and its efficiency is always less than 1. Hence the velocity ratio will always be greater than the mechanical advantage.

6.6 Law of a machine

If an experiment is carried out on a simple machine to determine the effort (E) required to raise a load (W) and a graph of E against W is plotted for a range of load values then a straight line graph, as illustrated in Fig. 6.1, would be obtained.

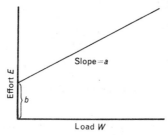

6.1 Effort–load graph for a simple machine

The relationship between effort E and the load W is therefore of the form:

$$E = aW + b$$

where a and b are constants, depending on the machine.

This equation is known as the *law of a machine*.

Putting $W = 0$ gives $E = b$. Thus, the constant b represents the effort necessary to overcome the frictional resistances at no load and is the intercept on the effort–load graph. The constant a is the slope of the effort–load graph and its value therefore depends upon the mechanical advantage of the machine.

6.7 Limiting efficiency of a machine

When the load on a machine is increased from zero the efficiency also increases. A typical graph of load (W) against efficiency (η) is shown in Fig. 6.2.

When the load on the machine is small the portion of the effort required to overcome frictional effects is large. This results in a low efficiency at low loads. However, as the load is increased the portion of the effort required to overcome friction becomes correspondingly smaller and so the efficiency increases. This increase in efficiency with load does not continue indefinitely, or we would attain the position of having a greater work output than work input in a given time, and a

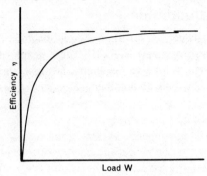

6.2 Typical load–efficiency graph for a simple machine

limiting efficiency is eventually reached. The limiting efficiency of a particular machine can be determined theoretically as follows. Now,

$$\text{Mechanical advantage} = \frac{\text{load }(W)}{\text{effort }(E)}$$

But from the law of a machine

$$E = aW + b$$

$$\text{M.A.} = \frac{W}{aW + b}$$

$$= \frac{1}{(a + b/W)}$$

As

$$\text{Efficiency} = \frac{\text{M.A.}}{\text{V.R.}}$$

it follows that, as the velocity ratio is constant for a particular machine, the maximum efficiency will occur when the mechanical advantage is a maximum. From the expression above for mechanical advantage it will be seen that as the load W increases so b/W decreases and the mechanical advantage increases. When b/W becomes so small that it can be neglected then the mechanical advantage reaches its maximum value of $1/a$. Hence

$$\text{Limiting efficiency} = \frac{1/a}{\text{V.R.}}$$

$$= \frac{1}{a \times \text{V.R.}}$$

6.8 Overhauling

A machine is said to overhaul when, in the absence of any effort, the load on the machine causes a reversal of action of the machine, i.e. the load overcomes the frictional forces. Such a situation is important, since if a machine overhauled upon removal of the effort the consequences could be serious. For example, a car jack which was capable of overhauling would make it impossible for one person to use and, although it is not a recommended practice, many people lie under a car which is jacked up. If the jack overhauled serious injury would result and we must therefore investigate the conditions under which overhauling could occur. Now,

$$\text{Efficiency } \eta = \frac{\text{work output}}{\text{work input}}$$

For a machine having an efficiency of 100% the work output would equal the work input. Thus,

$$\eta = \frac{\text{work input for machine of 100\% efficiency}}{\text{actual work input}}$$

$$= \frac{\text{effort for machine of 100\% efficiency}}{\text{actual effort}}$$

since the distance moved is constant. But for an efficiency of 100%, i.e. for an ideal machine

$$\text{Ideal effort} = \frac{\text{load}}{\text{M.A.}} = \frac{W}{\text{M.A.}}$$

For an efficiency of less than 100%:

Total load to be overcome by effort = applied load + frictional load

$$= W + F$$

$$\therefore \text{Actual effort} = \frac{W+F}{\text{M.A.}}$$

and

$$\eta = \frac{W/\text{M.A.}}{(W+F)/\text{M.A.}} = \frac{W}{W+F}$$

If overhauling is to occur when the effort is removed the load W must overcome the frictional resistance F, i.e.

$$W \geqslant F$$

Mechanical engineering science

In the particular case when $W = F$

$$\eta = \frac{W}{2W} = \tfrac{1}{2} \text{ or } 50\%$$

If $W > F$ then

$$W + F < 2W$$

and therefore

$$\eta > \frac{W}{2W}$$

i.e.

$$\eta > \tfrac{1}{2} \text{ or } 50\%$$

Thus, overhauling of a machine can only occur when its efficiency is greater than 50%.

6.9 Examples of simple machines

The simple wheel and axle, differential wheel and axle, pulley block, Weston differential wheel and axle, screw jack, etc. are all examples of simple machines. Problems involving these machines usually require the determination of the velocity ratio. This is obtained from the geometry of the machine, as illustrated in the worked examples which follow. In all problems it is usual to assume that no slipping occurs between any parts and also that there is no stretch of any chains or ropes.

Factors which must be considered in the design of a machine are the effect on the mechanical advantage and efficiency of any alteration to the velocity ratio. The velocity ratio is constant for a given machine and can only be altered by changing the machine geometry, i.e. by changing the size of a pulley, axle, gear, etc. Now, for a given operating efficiency any increase in the velocity ratio will result in an increase in the mechanical advantage, so enabling a greater load to be lifted by a certain effort. However, for a given loading on a machine the efficiency decreases as the velocity ratio increases. Hence the importance of carefully considering these factors in design.

6.10 Simple wheel and axle

This consists of a wheel and axle which are mounted together and rotate on the same axis, see Fig. 6.3. The effort is applied to a rope passing around the wheel while the load is raised by another rope winding onto the smaller axle.

Example 6.1

A wheel and axle has a 0.25 m diameter wheel and a 0.05 m diameter axle. An effort of 120 N is required to lift a load of 450 N. Calculate:
(a) the velocity ratio;
(b) the mechanical advantage;
(c) the efficiency.

Solution

A diagram is given in Fig. 6.3.

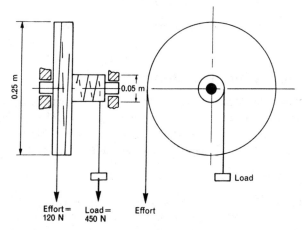

6.3 Simple wheel and axle

(a) Now, Velocity ratio $= \dfrac{\text{distance moved by effort}}{\text{distance moved by load}}$

Consider one revolution of the wheel and axle.

Distance through which = circumference of wheel
effort moves
$= \pi \times 0.25$ m

Distance through which = circumference of axle
load is raised
$= \pi \times 0.05$ m

Then, V.R. $= \dfrac{\pi \times 0.25}{\pi \times 0.05}$

$= 5$

100   Mechanical engineering science

(b) $\quad$ Mechanical advantage $= \dfrac{\text{load}}{\text{effort}}$
(for given load)

$$= \frac{450}{120} = 3.75$$

(c) $\quad$ Percentage efficiency $= \dfrac{\text{mechanical advantage}}{\text{velocity ratio}} \times 100$

$$= \frac{3.75}{5} \times 100$$

$$= 75\%$$

The velocity ratio is 5, the mechanical advantage 3.75 and the efficiency 75%.

6.11 Pulley block systems

The pulley block system is frequently used as a means of raising a load. The velocity ratio can be varied by adjusting the number of pulleys on the two blocks but, in any case, this ratio is never very high. A diagram of a simple pulley-block system is shown in Fig. 6.5.

Example 6.2

A pulley block system consists of three pulleys on the upper block and two on the lower block. A test on this lifting system gave the following results:

Load (N)	200	500	1000	1400	2000
Effort (N)	160	250	360	450	600

(a) Plot a graph of these results and hence determine the law of the machine.
(b) Plot a load-efficiency curve.
(c) Calculate the limiting efficiency of the pulley system.

Solution

(a) A graph of effort against load is plotted in Fig. 6.4.
The graph shows a linear relationship between the effort and the load.

$$\text{Intercept on effort axis} = b = 120$$

$$\text{Slope of graph} = a = \frac{480}{2000} = 0.24$$

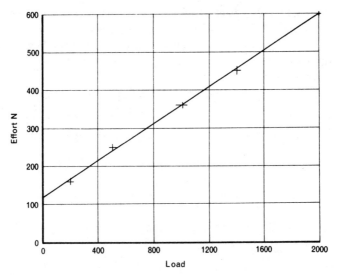

6.4 Effort–load graph—Example 6.2

The law of the machine is therefore

$$E = 0.24W + 120$$

(b) Referring to the diagram of the pulley block system, Fig. 6.5, if the load were raised by 1 m each length of chain between the pulley

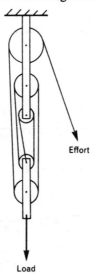

6.5 Pulley block diagram

blocks would have a slackness of 1 m. Thus, as there are five lengths of chain between the two blocks, the effort must move through a distance of 5 m in order to keep the chains taut. Hence, the velocity ratio = 5.

Alternatively, the velocity ratio can be obtained by simply counting the number of pulleys.

Then, from the load–effort results the following values are obtained.

Load [N]	200	500	1000	1400	2000
Effort [N]	160	250	360	450	600
M.A. = $\dfrac{\text{load}}{\text{effort}}$	1.25	2.0	2.78	3.11	3.34
$\eta = \dfrac{\text{M.A.}}{\text{V.R.}} \times 100\%$	25	40	55.6	62.2	66.8

The load-efficiency graph is plotted in Fig. 6.6.

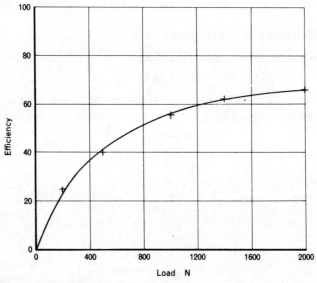

6.6 Efficiency–load graph—Example 6.2

(c) From Section 6.7:

$$\text{Limiting efficiency} = \frac{1}{a \times \text{V.R.}}$$

$$\frac{1}{0.24 \times 5}$$

0.834 or 83.4%

Simple machines 103

6.12 The screw-jack

The screw-jack is a simple device, making use of a screw thread, for raising relatively large loads by means of a small effort. The velocity ratio of a screw-jack is usually fairly high and although this tends to reduce the efficiency it is compatible with the design requirements because the efficiency must be less than 50% to prevent overhauling. A diagram of a simple screw-jack is given in Fig. 6.7.

Example 6.3

A screw-jack having a thread of 10 mm pitch is used to raise a load of 8750 N. If the efficiency of the screw-jack at this loading is 36%, calculate the effort that must be applied at a radius of 0.2 m.

Solution

A diagram is given in Fig. 6.7.

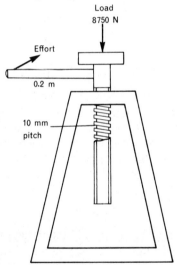

6.7 Simple screw-jack—Example 6.3

When the screw moves through one revolution the effort moves through a distance of $2\pi \times 0.2$ m, while the load is raised by 10 mm. Then,

$$\text{Velocity ratio} = \frac{\text{distance moved by effort}}{\text{distance moved by load}}$$

$$= \frac{2\pi \times 0.2}{10 \times 10^{-3}} = 40\pi$$

104 Mechanical engineering science

Now, Percentage efficiency = $\dfrac{\text{M.A.}}{\text{V.R.}} \times 100$

i.e. $36 = \dfrac{\text{M.A.}}{40\pi} \times 100$

$$\text{M.A.} = \dfrac{36 \times 40\pi}{100}$$

$$= 45.2$$

But, $\text{M.A.} = \dfrac{\text{load}}{\text{effort}}$

$\therefore 45.2 = \dfrac{8750}{E}$

$$E = \dfrac{8750}{45.2} = 193.5 \text{ N}$$

The effort that must be applied is 193.5 N.

6.13 The differential wheel and axle

This is similar to the wheel and axle but has two axles on the same axis as the wheel, as shown in Fig. 6.8. The effort is applied to the circumference of the wheel and the load is applied to a single pulley block suspended by a rope from the two axles. The rope is wound onto the two axles in opposite directions and this produces a much higher velocity ratio than can be obtained with the simple wheel and axle.

Example 6.4

A differential wheel and axle has a wheel of diameter 0.35 m and axles of diameters 0.08 m and 0.055 m respectively. Determine:

(a) the velocity ratio of the machine;
(b) the effort required to raise a load of 1200 N if the efficiency at this loading is 56%;
(c) the work done in raising the load through a height of 4 m.

Solution

A diagram of a differential wheel and axle is given in Fig. 6.8, with the dimensions appropriate to the above question.

Simple machines   105

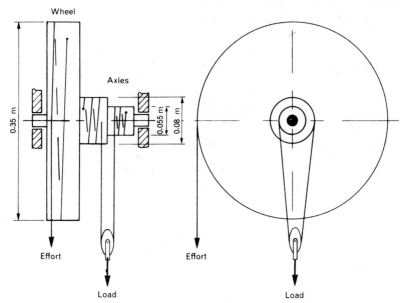

6.8 Differential wheel and axle—Example 6.4

(a) Consider one complete revolution of the wheel and axles.
Distance moved by effort = 0.35π m
Length of rope unwound from 0.055 m axle = 0.055π m
Length of rope wound onto 0.08 m axle   = 0.08π m
∴ Shortening of rope carrying load   = $0.08\pi - 0.055\pi$
                                         = 0.025π m

As this shortening is equally shared between the two sections of the rope, then

$$\text{distance through which load is raised} = \frac{0.025\pi}{2} = 0.0125\pi \text{ m}$$

But,   Velocity ratio = $\dfrac{\text{distance moved by effort}}{\text{distance moved by load}}$

$$= \frac{0.35\pi}{0.0125\pi} = 28$$

(b) Now, $\eta = \dfrac{\text{M.A.}}{\text{V.R.}}$

∴ $0.56 = \dfrac{\text{M.A.}}{28}$

106 Mechanical engineering science

$$M.A. = 28 \times 0.56 = 15.68$$

But, $$M.A. = \frac{load}{effort}$$

$$\therefore Effort = \frac{load}{M.A.} = \frac{1200}{15.68} = 76.5 N$$

(c) When the load is raised through a distance of 4 m;

$$Work\ output = 1200\ [N] \times 4\ [m]$$
$$= 4800\ J$$

But, $$Work\ input = \frac{work\ output}{efficiency}$$

$$= \frac{4800}{0.56}$$

$$= 8570\ J$$

The velocity ratio is 28, the effort required to raise a load of 1200 N is 76.5 N and the work done to raise this load through 4 m is 8570 J.

6.14 Weston differential pulley block

This uses three pulleys, as shown in Fig. 6.9, and by varying the size of the pulleys on the upper block the velocity ratio can be altered to suit requirements. Ropes or chains are used and for very heavy loads link chains are frequently used, the pulleys then having flats or teeth.

Example 6.5

A Weston differential pulley block has 12 teeth on the large pulley and 11 teeth on the smaller pulley. If efforts of 75 N and 120 N are required to raise loads of 1100 N and 2000 N respectively, determine:

(a) the velocity ratio of the pulley block;
(b) the law of the machine;
(c) the efficiency at each loading;
(d) the limiting efficiency.

Solution

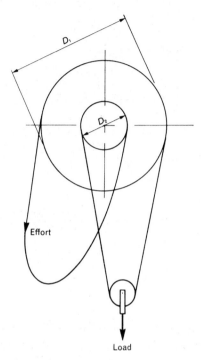

6.9 Weston Differential Pulley block—Example 6.5

The velocity ratio is given by:

$$\text{V.R.} = \frac{2D_1}{D_1 - D_2} \quad \text{or} \quad \frac{2n_1}{n_1 - n_2}$$

where D_1 = diameter of larger pulley
or n_1 = number of teeth or flats on larger pulley
and D_2 = diameter of smaller pulley
or n_2 = number of teeth or flats on smaller pulley.

The proof of these relationships is obtained in a similar manner to the derivation of the velocity ratio in Example 6.4 and is therefore left for proof as part of Problem 15.

(a) Using
$$\text{V.R.} = \frac{2n_1}{n_1 - n_2}$$

gives
$$\text{V.R.} = \frac{2 \times 12}{12 - 11} = 24$$

The velocity ratio of the machine is 24.

108   Mechanical engineering science

(b) Assuming that the law of the machine can be written in the form

$$E = aW + b$$

we obtain for the two loadings

$$120 = 2000a + b \qquad (1)$$

and

$$75 = 1100a + b \qquad (2)$$

By subtraction

$$45 = 900a$$

$$a = 0.05$$

In equation (1)

$$120 = 2000 \times 0.05 + b$$

$$= 100 + b$$

$$\therefore b = 20$$

The law of the machine is therefore $E = 0.05W + 20$.

(c) Since

$$\text{M.A.} = \frac{\text{load}}{\text{effort}}$$

For 1st load:

$$\text{M.A.} = \frac{1100}{75} = 14.68$$

Then

$$\eta = \frac{\text{M.A.}}{\text{V.R.}} = \frac{14.68}{24} = 0.612 \text{ or } 61.2\%$$

For 2nd load:

$$\text{M.A.} = \frac{2000}{120} = 16.66$$

$$= \frac{16.66}{24} = 0.69 \text{ or } 69\%$$

The efficiency of the pulley block when a load of 1100 N is raised is 61.2% and when the load is 2000 N the efficiency is 69%.

(d) From Section 6.7:

$$\text{Limiting efficiency} = \frac{1}{a \times \text{V.R.}}$$

$$= \frac{1}{0.05 \times 24}$$

$$= 0.833 \text{ or } 83.3\%$$

The limiting efficiency of the pulley block is 83.3%.

Simple machines

Problems

1. A wheel and axle is used to raise a load of 850 N through a height of 0.6 m. If the axle diameter is 60 mm and the wheel diameter is 350 mm what effort will be required if the efficiency at this loading is 70%?

2. A screw-jack has a thread of 8 mm pitch. What is the maximum load that can be raised by an effort of 300 N applied at a radius of 0.18 m if the efficiency is 45%?

3. A simple wheel and axle has a wheel of 0.34 m diameter and an axle of 0.04 m diameter. What effort will be required to raise a load of 1500 N? Assume an efficiency of 65%.

How much work is expended in overcoming friction when the load is raised by 0.4 m?

4. In a laboratory test on a screw-jack the following results were obtained:

Load (N)	180	250	320	390
Effort (N)	11.8	15.4	18.9	22.4

The effort is applied at the circumference of a pulley 0.20 m diameter. The pitch of the screw is 12 mm single start:

(*a*) calculate the velocity ratio;
(*b*) plot the load/efficiency curve;
(*c*) obtain the law of the machine;
(*d*) calculate the limiting efficiency.

5. A screw-jack has a 10 mm pitch thread and the effort is applied at a radius of 0.14 m. If a load of 5000 N is raised by means of an effort of 290 N determine the efficiency of the screw-jack.

6. In a test on a simple lifting machine it was found that an effort of 65 N was required to raise a load of 1080 N and that the efficiency under these conditions was $41\frac{2}{3}$%. Calculate:

(*a*) the velocity ratio;
(*b*) the work done by the effort in raising the load 0.6 m;
(*c*) the effort that would be required to lift the same load if the efficiency is raised to 52% by lubricating the equipment.

7. A screw-jack has a screw of 12.5 mm pitch and the effort is applied at the end of an arm 0.375 m long. The effort P required to raise a load W is given by

$$P = \frac{W}{130} + 67$$

where P and W are in newtons.

Calculate the efficiency of the screw-jack when the load lifted is (a) 6500 N, (b) 9000 N.
Why does the efficiency vary with the load being lifted?

8. If in Problem 7 the pitch is changed to 7.5 mm and the arm length to 0.28 m what would be the efficiency when the load being lifted is (a) 10 kN, (b) 18 kN?

9. A simple machine has a velocity ratio of 60. In a test on the machine it was found that an effort of 19.6 N was required to raise a load of 240 N and an effort of 52 N was required to raise a load of 960 N. Determine the law of the machine and calculate the efficiency of the machine for a load of 1440 N.

10. Derive an expression for the limiting efficiency of a simple machine. Assume the law of the machine to be of a linear form.

A load of 800 N is raised by means of a wheel and axle having a 0.20 m diameter wheel and a 0.05 m diameter axle. If the effort required is 280 N calculate the efficiency at this loading.

11. A lifting gear is used to raise a load of 6850 N. The velocity ratio is 36 and the efficiency at this loading is 30%. In a no load test on the lifting gear it was found that an effort of 42 N was necessary to overcome friction. Obtain the law of the machine and calculate:

(a) the effort required to raise a load of 14 500 N;
(b) the limiting efficiency of the machine.

12. (a) Show that the velocity ratio of a differential wheel and axle is given by

$$\text{Velocity ratio} = \frac{2D}{d_2 - d_1}$$

Where D is the diameter of the wheel and d_1 and d_2 are the diameters of the axles.

What is the principal advantage of a differential wheel and axle as compared with a simple wheel and axle?

(b) A differential wheel and axle has an 0.44 m diameter wheel and axles of 0.054 m and 0.080 m respectively. What load can be raised by an effort of 250 N if the efficiency under these conditions is 40%? How much work is done against friction when the load is raised through a height of 0.35 m?

13. The axles of a differential wheel and axle have diameters of 0.05 m and 0.07 m respectively and the wheel is 0.40 m diameter. It is found

Simple machines   111

during a test that an effort of 300 N is required to lift a load of 4800 N. Calculate the efficiency of the wheel and axle at this loading.

14. A screw-jack has a two start thread with a pitch of 6 mm. Determine the force required at the end of a jack handle, 0.22 m long, to raise a load of 15 000 N if the efficiency at this load is 28%.

15. Derive an expression for the velocity ratio of a Weston differential pulley block.

A differential block has 10 flats on the large pulley and 9 on the small pulley. Determine the effort required to raise a load of 1250 N if the efficiency at this load is 35%.

16. State and explain the condition under which a simple lifting machine will overhaul, i.e. reverse under load.

A Weston differential block has pulleys of 0.27 m and 0.235 m diameter respectively. The following results were obtained during a test.

$$\text{Load [N]} \quad 400 \quad 1440$$
$$\text{Effort [N]} \quad 64.8 \quad 148$$

Determine the law of the machine, assuming it to be linear. Hence obtain the effort required to raise a load of 5000 N, and also the limiting efficiency of the machine.

At what load will the machine first overhaul?

7

Kinematics—velocity and acceleration

7.1 Kinematics

Kinematics deals with the motion of bodies without considering the forces which produce such motion.

7.2 Distance and displacement

When a body moves, so changing its position, the distance it has moved is measured by the length of its path of motion. Distance is given the symbol s and is measured in metres (m).

If the path curves, the length of the straight line between the starting and finishing points of the motion is called the displacement. Thus, in Fig. 7.1 the length of the dotted line is the distance along a possible path between A and B, while the continuous line is the displacement between A and B.

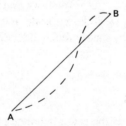

7.1 Distance and displacement

When referring to distance, only length is involved and therefore this is a scalar quantity. However, to specify the displacement between two points both the length (or distance) and direction must be known, and therefore displacement is a vector quantity. Thus, if we say that the distance between two points A and B is 10 km then this is a scalar quantity but if we say that B is 10 km north-east of A then this is a vector quantity, since both distance and direction are specified.

7.3 Speed

Speed is defined as the rate of change of distance with time and as no direction is involved speed is a scalar quantity.

Thus, if a body travels through a distance of 2 m in 4 s its average speed is 2/4 or 0.5 m/s. This in no way implies that the speed of the body remained constant throughout the motion, merely that for the time interval considered the average speed was 0.5 m/s.

$$\text{Average speed} = \frac{\text{distance travelled}}{\text{time taken}}$$

$$v_{av} = \frac{s}{t}$$

The speed of a body is constant if equal distances are covered in equal periods of time, irrespective of the time interval chosen, i.e. v_{av} is the same for any period of time t.

7.4 Velocity

The velocity of a body is its speed measured in a definite direction. Thus, speed and velocity are equal in magnitude but whereas speed, having only magnitude, is a scalar quantity, velocity, having both magnitude and direction, is a vector quantity and can therefore be represented by a line drawn to a chosen scale and in a definite direction. Velocity is represented by the symbol v, its basic unit being metres/second (m/s).

7.5 Acceleration

Acceleration is defined as the rate of change of velocity with time. As velocity is specified by both magnitude (speed) and direction, it follows that if there is a change in either the speed of a body or its direction of motion then the body is subject to an acceleration. This can be illustrated as follows. Consider a car travelling around a circular track at a constant speed. Because the direction of motion of the car is continually changing it follows that the velocity, and hence the acceleration,

114 Mechanical engineering science

are also continually changing. This is due solely to change in direction, the magnitude of both the velocity and acceleration remaining constant throughout.

When the speed of a body decreases with time it is said to decelerate, or retard, and the rate of decrease of speed, which is in fact a negative acceleration, is known as the deceleration or retardation.

$$\text{Average acceleration} = \frac{\text{change in velocity}}{\text{time taken}}$$

Acceleration is represented by the symbol a and its basic units are metres/second^2 (m/s^2).

7.6 Distance–time graph

(*a*) CONSTANT SPEED

Consider a body which moves with a constant speed. This means that the body will travel equal distances in equal periods of time, i.e. a car travelling at a constant speed of 20 metres/second (m/s) will have travelled 20 m after 1 s, 40 m after 2 s, etc. Thus,

$$\text{Distance travelled} = \text{speed} \times \text{time}$$

$$s = vt$$

A graph of distance against time is therefore a straight line, as shown in Fig. 7.2. The slope of this graph is equal to the speed.

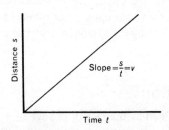

7.2 Distance–time graph for constant speed

(*b*) VARYING SPEED

If the speed of a body is continually varying then the distance–time will no longer be a straight line. Such a graph is shown in Fig. 7.3.

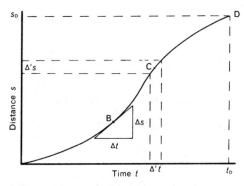

7.3 Distance–time graph for varying speed

The relationship proved for constant speed, viz. the speed is equal to the slope of the distance–time graph, is still true. Thus at point B on Fig. 7.3.

Speed = slope of distance–time graph

$$v_b = \frac{\Delta s}{\Delta t}$$

Another method of determining the speed at a stage of the journey is illustrated at C. A short length of the graph, to either side of C, is used to obtain $\Delta' s$ and $\Delta' t$ as shown.

$$v_c = \frac{\Delta' s}{\Delta' t}$$

For the complete journey

$$\text{Average speed} = \frac{\text{total distance travelled}}{\text{time taken}}$$

$$v_{av} = \frac{s_D}{t_D}$$

7.7 Speed–time graph

(a) CONSTANT SPEED

If a body travels at a constant speed then the speed–time graph will be as shown in Fig. 7.4.

Now, from Section 7.6:

Distance travelled = speed × time
= area under speed–time graph (shaded)

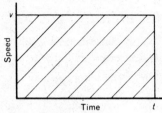
7.4 Speed–time graph for constant speed

(b) VARYING SPEED

The result obtained above is also applicable to a body whose speed varies with time.

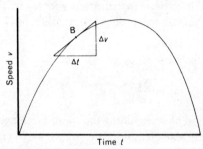
7.5 Speed–time graph for varying speed

Referring to Fig. 7.5 the distance travelled is equal to the area under the graph. The numerical calculation of such an area is illustrated by Example 7.4.

From Section 7.5.

$$\text{Acceleration} = \frac{\text{change in velocity}}{\text{time taken}}$$

For the speed–time graph of Fig. 7.5 it follows that as the speed is continually changing then so the acceleration is also continually changing. At any instant, however, the acceleration will be equal to the slope of the graph.

Thus, at B

$$\text{Acceleration} = \text{slope of speed–time graph}$$

$$= \frac{\Delta v}{\Delta t}$$

7.8 Constant (or uniform) acceleration

In this book we are mainly concerned with the motion of bodies having a uniform (constant) acceleration. Thus, from Section 7.7, it follows that the slope of the velocity–time graph will be constant and equal to the acceleration.

Let us consider a body which starts with an initial velocity v_1 and accelerates uniformly to a velocity v_2 in a period of time t. The speed–time graph for such a motion is given in Fig. 7.6.

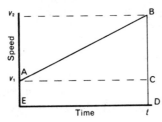

7.6 Speed–time graph for constant acceleration

From Section 7.7:

$$\text{Acceleration} = \text{slope of speed–time graph}$$
$$= \frac{BC}{AC}$$
$$\therefore a = \frac{v_2 - v_1}{t}$$

or
$$v_2 = v_1 + at \tag{7.1}$$

Also,

$$\text{Distance travelled} = \text{area under velocity–time graph}$$
$$= ACDE + ABC$$
$$s = v_1 t + \tfrac{1}{2}(v_2 - v_1)t$$
$$s = \left(\frac{v_1 + v_2}{2}\right)t \tag{7.2}$$

Substituting from (7.1) into (7.2) for v_2

$$s = \left(\frac{v_1 + v_1 + at}{2}\right)t$$
$$s = v_1 t + \tfrac{1}{2}at^2 \tag{7.3}$$

118   Mechanical engineering science

Squaring equation (7.1)

$$v_2^2 = (v_1+at)^2$$
$$= v_1^2 + 2v_1at + a^2t^2$$
$$= v_1^2 + 2a(v_1t + \tfrac{1}{2}at^2)$$
$$= v_1^2 + 2as \qquad (7.4)$$

Substituting for v_1 from (7.1) into (7.2)

$$s = \left(\frac{v_2 - at + v_2}{2}\right)t$$
$$s = v_2 t - \tfrac{1}{2}at^2 \qquad (7.5)$$

The equations (7.1) to (7.5) are the equations of linear motion for uniform acceleration. They are important and should be learned. For convenience they are summarised together:

$$v_2 = v_1 + at \qquad v_1 = \text{initial velocity}$$
$$s = \left(\frac{v_1 + v_2}{2}\right) \times t \qquad v_2 = \text{final velocity}$$
$$s = v_1 t + \tfrac{1}{2}at^2 \qquad a = \text{acceleration}$$
$$v_2^2 = v_1^2 + 2as \qquad s = \text{distance travelled}$$
$$s = v_2 t - \tfrac{1}{2}at^2 \qquad t = \text{time for velocity change to occur}$$

Note that there are five equations covering five quantities but each involving only four of the five quantities. Also, the equations do not apply if the acceleration is not uniform.

Example 7.1

A train starts from rest and accelerates uniformly to a speed of 50 km/h in 15 s. Determine:

(a) the acceleration of the train;
(b) the distance travelled in this time.

Solution

(a) $\qquad 50 \text{ km/h} = \dfrac{50 \times 1000}{60 \times 60} = 13.88 \text{ m/s}$

Kinematics—velocity and acceleration

Applying equation (7.1)

$$v_2 = v_1 + at$$

where $v_1 = 0$; $v_2 = 13.88$ m/s; $t = 15$ s, then,

$$13.88 \text{ [m/s]} = a \times 15 \text{ [s]}$$

$$a = \frac{13.88}{15} = 0.925 \text{ m/s}^2$$

(b) Applying equation (7.2)

$$s = \left(\frac{v_1 + v_2}{2}\right) t$$

$$= \left(\frac{0 + 13.88}{2}\right) \times 15$$

$$= 104.1 \text{ m}$$

The acceleration is 0.925 m/s² and the distance travelled is 104.1 m.

Example 7.2

A car starts from rest and accelerates uniformly for a period of 12 s. It then travels at a constant speed for the next 8 min after which it comes to rest in a further 15 s. The total distance travelled by the car is 3.5 km. Determine:

(a) the constant speed;
(b) the acceleration.

Draw a velocity–time graph for the journey.

Solution

The journey can be divided into three stages:

(1) an acceleration stage;
(2) a constant speed stage;
(3) a deceleration stage.

Let the constant velocity be V [m/s] and the acceleration be a [m/s²].

(a) The constant speed

    (1) For acceleration stage:

Applying $\quad s = \left(\frac{v_1 + v_2}{2}\right) \times t$

distance travelled,
$$s_1 = \left(\frac{0+V}{2}\right)12$$
$$= 6V \text{ m}$$

(2) For constant speed stage:
distance travelled,
$$s_2 = \text{speed} \times \text{time}$$
$$= V \times 8 \times 60$$
$$= 480V \text{ m}$$

(3) For deceleration stage:
Applying $\quad s = \left(\frac{v_1+v_2}{2}\right) \times t$

distance travelled,
$$s_3 = \left(\frac{V+0}{2}\right) \times 15$$
$$= 7.5V \text{ m}$$

Now,
Total distance travelled $= s_1 + s_2 + s_3$
$$3.5 \times 1000 = 6V + 480V + 7.5V$$
$$V = \frac{3.5 \times 1000}{493.5}$$
$$= 7.1 \text{ m/s}$$
$$= 25.56 \text{ km/h}$$

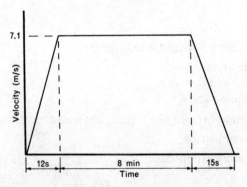

7.7 Velocity–time graph—Example 7.2

(b) The acceleration:
Applying
$$v_2 = v_1 + at$$
$$7.1 = 0 + 12a$$
$$a = 0.591 \text{ m/s}^2$$

The constant speed is 25.56 km/h and the acceleration is 0.591 m/s². A velocity-time graph for the journey is drawn as Fig. 7.7.

Example 7.3

A train starts from a station and accelerates uniformly to a speed of 110 km/h in a time of 2 min. It then travels at constant speed for the next 32 km and is brought to rest at the next station with a uniform deceleration of 0.62 m/s². Draw a velocity-time graph for the journey and determine from it:

(a) the total time of the journey;
(b) the distance between the stations.

Solution

$$110 \text{ km/h} = \frac{110 \times 1000}{60 \times 60} \text{ m/s}$$
$$= 30.5 \text{ m/s}$$

A velocity-time graph is drawn in Fig. 7.8. The journey can be divided into three stages:

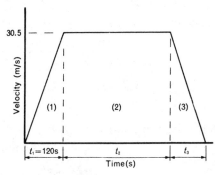

7.8 Velocity-time graph—Example 7.3

(1) acceleration to 30.5 m/s in 120 s;
(2) constant velocity of 30.5 m/s for a time of t_2 [seconds] while a distance of 32 km is travelled;
(3) deceleration at 0.62 m/s² for time t_3 [seconds].

To solve the problem we require the time and distance travelled for each of the three stages.

(1) Acceleration:

$$\text{Period of acceleration } (t_1) = 120 \text{ s}$$

Distance travelled,

$$\begin{aligned} s_1 &= \text{area (1) under velocity–time graph} \\ &= \tfrac{1}{2} \times 120 \text{ [s]} \times 30.5 \text{ [m/s]} \\ &= 1830 \text{ m} \\ &= 1.83 \text{ km} \end{aligned}$$

(2) Constant velocity:

Distance travelled, $\quad s_2 = 32$ km

But, Distance travelled = area (2) under velocity–time graph

$$\therefore 32 \times 1000 \text{ [m]} = 30.5 \text{ [m/s]} \times t_2$$

$$t_2 = \frac{32 \times 1000}{30.5} \text{ s}$$

$$= 1050 \text{ s}$$

(3) Deceleration:

Rate of deceleration = slope of velocity–time graph

$$0.62 = \frac{30.5 - 0}{t_3}$$

$$t_3 = 49.2 \text{ s}$$

Distance travelled = area (3) under velocity–time graph

$$\begin{aligned} s_3 &= \tfrac{1}{2} \times 49.2 \text{ [s]} \times 30.5 \text{ [m/s]} \\ &= 750 \text{ m} \\ &= 0.75 \text{ km} \end{aligned}$$

Then,

(a) $\quad$ Total time of journey $= t_1 + t_2 + t_3$

$$= 120 + 1050 + 49.2 \text{ s}$$
$$= 1219.2 \text{ s}$$
$$= 20 \text{ min } 19.2 \text{ s}$$

(b) $\quad$ Distance between stations $= s_1 + s_2 + s_3$

$$= 1.83 + 32 + 0.75 \text{ km}$$
$$= 34.58 \text{ km}$$

Kinematics—velocity and acceleration

Example 7.4

The variation of the velocity of a body with time is given in the following table.

Velocity (m/s)	0	0.8	2	5.4	11.4	24
Time (s)	0	2	4	8	12	16

Draw the velocity–time graph and hence determine the distance travelled during the first 12 s of the motion.

Solution

The velocity–time graph is drawn in Fig. 7.9.

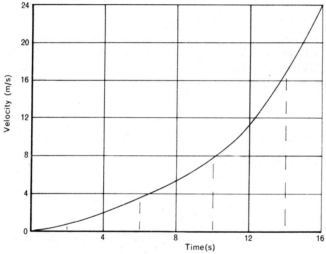

7.9 Velocity–time graph—Example 7.4

The distance travelled during the first 12 s of the motion is equal to the area under the graph up to this time. This area can be obtained by applying the mid-ordinate rule, taking 6 divisions, each of 2 seconds interval.

$$\text{Average velocity ordinate} = \frac{(0.4+1.2+2.6+4.4+6.6+9.6)}{6}$$

$$= 4.133 \text{ m/s}$$

Then, Distance travelled = average velocity × time
$$= 4.133 \times 12$$
$$= 49.6 \text{ m}$$

The distance travelled during the first 12 s is 49.6 m.

7.9 Motion under gravity

If a body falls freely under gravity, with the air resistance being negligible, then it has been found that the acceleration is uniform. This acceleration varies over the earth's surface but a value of 9.81 m/s² is normally used.

It must be emphasised that most bodies in free fall rapidly attain a velocity at which the air resistance is appreciable and the acceleration is therefore considerably reduced. In fact, the terminal velocity for most bodies is approximately 200 km/h. This limiting velocity is attained when the air resistance, or drag, which is generally a function of the velocity, is equal to the weight of the body.

7.10 Resolution of velocities

In the same way that a force, being a vector quantity, can be resolved into two components having the same vectorial sum, so a velocity, also being a vector, can likewise be resolved. Thus, in Fig. 7.10 *ab* can be resolved into components *ac* and *cb*.

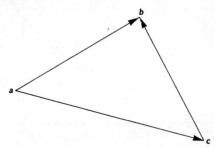

7.10 Resolution of velocities

However, in the same manner as for forces, it is usually most convenient to resolve a velocity into two directions at right angles, i.e. into rectangular components. Thus, if a body has a velocity V at an angle θ to the horizontal this can be resolved into a horizontal component $V \cos \theta$, and a vertical component $V \sin \theta$ as shown in Fig. 7.11.

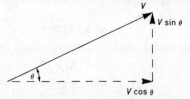

7.11 Resolution of a velocity into rectangular components

An acceleration also being a vector quantity, can be resolved in a similar manner.

Example 7.5

A ball is projected upwards at an angle of 40° to the horizontal from the top of a tower which is 44 m above the level ground. If the initial velocity of the ball is 25 m/s and air resistance is neglected, calculate:

(a) the time of flight;
(b) the horizontal range;
(c) the magnitude and direction of the velocity of the ball when it hits the ground.

Solution

A diagram is given in Fig. 7.12.

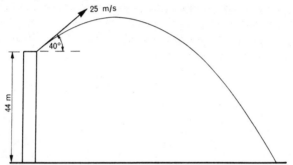

7.12 Example 7.5

Horizontal component of projection velocity = 25 cos 40°
 = 19.15 m/s
Vertical component of projection velocity = 25 sin 40°
 = 16.05 m/s

(a) Now, when the flight of the ball is completed it is at ground level and is therefore 44 m lower than when it was initially fired from the top of the tower. Thus, as our positive direction is upwards, i.e. in the direction of the initial vertical velocity of the ball, the effective vertical distance travelled is −44 m.

As the horizontal component of the projection velocity is not affected by gravity, i.e. the horizontal component remains constant throughout the flight, we can apply

$$s = v_1 t + \tfrac{1}{2}at^2$$

to the vertical motion of the complete flight.

$s = -44$ m; $v_1 = 16.05$ m/s; $a = -9.81$ m/s² (acceleration is downwards)

Then
$$-44 = 16.05t + \tfrac{1}{2}(-9.81)t^2$$
$$4.905\, t^2 - 16.05t - 44 = 0$$
$$t = \frac{16.05 \pm \sqrt{16.05^2 + 4 \times 4.905 \times 44}}{2 \times 4.905}$$
$$= +5.05 \text{ s} \quad \text{or} \quad -1.775 \text{ s}$$

The negative value for t is rejected.

(b) Horizontal range = horizontal component of velocity × time
$$= 19.15 \text{ [m/s]} \times 5.05 \text{ [s]}$$
$$= 96.7 \text{ m}$$

(c) When the ball hits the ground the horizontal component of its velocity will still be 19.15 m/s.

The vertical component of its velocity can be obtained by applying
$$s = \left(\frac{v_1 + v_2}{2}\right)t$$
to the complete vertical flight. Then,
$$-44 = \left(\frac{16.05 + v_2}{2}\right) \times 5.05$$
$$v_2 = -\frac{88}{5.05} - 16.05$$
$$= -33.45 \text{ m/s, i.e. downwards}$$

These component velocities are shown in Fig. 7.13, where V is the

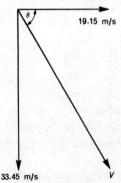

7.13 Component velocities of ball prior to impact—Example 7.5

Kinematics—velocity and acceleration

resultant velocity acting at angle θ to the horizontal. Then

$$V^2 = 19.15^2 + 33.45^2$$
$$V = 38.6 \text{ m/s}$$

and
$$\tan \theta = \frac{33.45}{19.15} = 1.747$$
$$\theta = 60° \, 12'$$

7.11 Relative velocity

When we say that a train is travelling at 80 km/h we really mean that it is travelling at 80 km/h relative to the earth's surface, which we think of as being at rest. When the velocity of a body is quoted relative to the earth it is known as the absolute velocity. In many instances the velocity of a body is quoted relative to that of another body, which is itself moving relative to the earth, and the term relative velocity is used. Thus the relative velocity of a body is the velocity with which it appears to be moving when viewed from another body. Let us consider a simple example.

Consider two cars on a straight road which runs east–west. Let us suppose that car A is travelling east at 60 km/h while car B is travelling west at 40 km/h, as shown in Fig. 7.14.

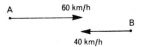

7.14 Relative velocity—Bodies moving along the same line

To a person in car A, car B appears to be travelling west at 100 km/h. i.e. the velocity of B relative to A is 100 km/h, due west. Conversely, to a person in car B, the apparent velocity of car A is 100 km/h, due east.

If the direction of the velocity of A is taken as positive then the velocity of B, being in the opposite direction, is -40 km/h. Then,

$$\text{Velocity of B relative to A} = \text{velocity of B} - \text{velocity of A}$$
$$= (-40) - (+60)$$
$$= -100 \text{ km/h}$$

the $-$ve sign indicating that the velocity of B relative to A is in a direction from east to west.

Velocity of A relative to B = velocity of A − velocity of B
= (+60) − (−40)
= +100 km/h

the +ve sign indicating that the velocity of A relative to B is in a direction from west to east.

It will be seen that the relative velocity between two bodies moving along the same line is obtained by simple algebraic subtraction.

We must now consider the case when the two bodies are not moving along the same line. This is shown in Fig. 7.15 where A and B are moving in the same plane with velocities v_A and v_B respectively.

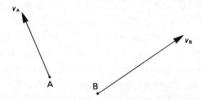

7.15 Relative velocity—Bodies not moving along the same line

To determine the velocity of B relative to A we must find the velocity with which B would appear to be moving if A were at rest. Thus, a velocity of $-v_A$ is applied to both bodies, as shown in Fig. 7.16(i). This effectively brings A to rest while the components of the velocity of B are its true velocity v_B and its superimposed velocity $-v_A$. The resultant velocity of B is given by vector *ab* in Fig. 7.16(ii).

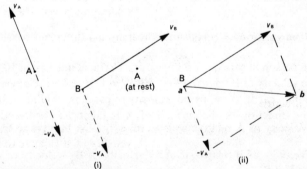

7.16 Velocity of B relative to A

Now, as A is apparently at rest, the vector *ab* represents the velocity of B as it appears to an observer on A and is therefore the velocity of B relative to A. This result was obtained by adding vectorially the velocity of A reversed to the velocity of B. Writing this as a vector equation:

Velocity of B relative to A = absolute velocity of B
− absolute velocity of A

If we had chosen to find the velocity of A relative to B we would have obtained a velocity of equal magnitude but in the opposite direction, i.e. vector **ba**.

If, instead of drawing the velocities as in Fig. 7.16(ii) we had drawn the absolute velocities of A and B relative to some arbitrary fixed point O, as shown in Fig. 7.17, then the relative velocity between the bodies is obtained in the same way as before.

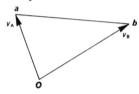

7.17 Velocity of B relative to A

From Fig. 7.17:

$$Ob = Oa + ab$$
$$ab = Ob - Oa$$

Or in words,

Velocity of B relative to A = velocity of B relative to O
− velocity of A relative to O

But, as vector **ab** = − vector **ba** the first equation can be written as

$$Ob = Oa - ba$$
$$ba = Oa - Ob$$

Or in words

Velocity of A relative to B = velocity of A relative to O
− velocity of B relative to O

Example 7.6

A ship A, steaming in a north-westerly direction, at 16 km/h observes a second ship B, 5 km due north. Ship B is steaming due south at 22 km/h. Determine the velocity of ship B relative to ship A and the shortest distance that separates the two.

Solution

The positions of the ships, and their individual velocities are indicated in Fig. 7.18.

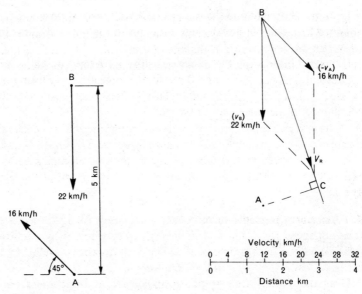

7.18 Space diagram 7.19 Velocity vector diagram
Example 7.6

To determine the velocity of B relative at A, a velocity of 16 km/h is assumed to be applied to both, in a direction such that A is apparently brought to rest.

The velocity of B relative to A is then obtained as indicated in Fig. 7.19 and from this diagram the relative velocity V_R is 35.2 km/h in a direction 19° east of south.

To determine the shortest distance that separates the two ships, ship A is imagined to be at rest and ship B to be moving with the relative velocity of 35.2 km/h along the line BC. The shortest distance separating the two ships is then given by the perpendicular AC drawn from A onto BC.

From Fig. 7.19 this distance is 1.61 km.

The velocity of ship B relative to ship A is 35.2 km/h in a direction 19° east of south and the shortest distance that separates the two ships is 1.61 km.

Problems

1. A vehicle starting from rest accelerates uniformly to a speed of 80 km/h in a time of 20 s. Determine the acceleration and the distance travelled in this time.

2. A body starts from rest and accelerates uniformly until it reaches a speed of 72 km/h. It travels at this speed for the next 4 min and is then uniformly brought to rest. If the rate of acceleration is twice the rate of retardation and the total distance travelled is 6.4 km, calculate:

(a) the acceleration;
(b) the total time of motion.

3. A train travelling at 90 km/h is checked by a signal. It decelerates uniformly over a distance of 200 m until its speed is 24 km/h. It travels 300 m at this speed and then accelerates back to its original speed at a rate of 0.55 m/s². Calculate the total time lost as a result of the signal check.

4. A car starts from rest and accelerates uniformly for 15 s, travels at a constant speed for 4 min and is then brought to rest with a uniform retardation in a further time of 18 s. The total distance travelled by the car is 7.2 km.

Draw the velocity-time graph for the journey and calculate from it:

(a) the maximum speed attained by the car;
(b) the distance travelled during the first minute of the journey.

5. The speed of a body varies with time as given in the following table.

Time (s)	0	2	4	6	8	10
Speed (m/s)	0	2.1	3.3	4.1	4.7	5.1
Time (s)	12	14	16	18	20	
Speed (m/s)	5.5	5.7	5.8	5.9	5.9	

By plotting a suitable graph determine the distance travelled by the body in (a) 10 s, (b) 20 s from rest.

6. The following table gives the variation of the speed of a car with time.

Speed (km/h)	0	7.5	14	20	20	11	0
Time (s)	0	3	6	9	12	15	18

Plot a graph and obtain the total distance travelled. What is the acceleration of the car after a time of 4 s?

132 Mechanical engineering science

7. The speed of a train on a journey between two stations varies as follows.

Time (min)	0	1	2	3	4	5	6	7	8	9
Speed (km/h)	0	30	50	65	75	78	75	65	40	0

Determine:

(a) the distance between the two stations;
(b) the acceleration of the train after 1 min;
(c) the retardation of the train after 8 min.

8. A mine cage starts from rest at the bottom of a mine shaft and is accelerated uniformly for a period of 12 s until a speed of 7.2 m/s is attained. The speed is maintained constant for the next 24 s and the cage is then brought to rest in a time of 9 s with a uniform retardation. Determine the depth of the mine shaft.

9. A vehicle accelerates uniformly from rest until it attains a speed of 100 km/h. It then travels at this speed for a certain period before decelerating uniformly to rest. The complete journey covers 7 km and takes 5 min 15 s. If the rate of retardation is twice the rate of acceleration, determine:

(a) the acceleration;
(b) the time for which the vehicle travels at a constant speed.

10. A shell is fired from a gun with a velocity of 250 m/s at an angle of 40° to the horizontal. Calculate the greatest height attained by the shell and its horizontal range. Neglect the effects of air resistance.

11. A marble is dropped from the window of a building so that it falls vertically. If the height of the window above the road surface is 80 m, calculate:

(a) the speed of the marble when it hits the road;
(b) the time taken.

12. A shell is projected at an angle of 36° upwards from a horizontal plane with a velocity of 460 m/s. Calculate:

(a) the range of the shell;
(b) the total time of flight;
(c) the maximum height reached by the shell.

13. A cricketer can throw a ball with a maximum velocity of 32 m/s. When fielding on the boundary he is at a distance of 67.5 m from the wicket-keeper. At what angle should he throw the ball in order that it

reaches the wicket-keeper in the shortest possible time? What is this time?

14. A rocket is fired from the ground at an angle of 50° to the horizontal and it accelerates uniformly, from rest, along this line of flight until it reaches a velocity of 3250 km/h after a time of 12 s. The propulsion unit is then cut out and free flight follows. Neglecting any effects due to air resistance, calculate:

(*a*) the maximum height reached;
(*b*) the total time of flight;
(*c*) the horizontal range.

15. A ship, A, steaming 20° north of east at 18 km/h sights another ship, B, 5 km due south. Ship B is steaming north-east at 14 km/h. Determine:

(*a*) the velocity of ship B relative to ship A;
(*b*) the shortest distance which separates the two ships.

16. An aeroplane flying due east at 900 km/h observes another aeroplane 35 km south-east. The second aeroplane is flying north-east at 600 km/h. What is the velocity of the second aircraft relative to the first? If the aeroplanes were at the same altitude what would be their closest distance of approach?

17. A steamship travelling due north at 36 km/h observes a helicopter that is apparently travelling north-west at a speed of 108 km/h. Determine the true magnitude and direction of the velocity of the helicopter.

18. At a certain time a ship A, travelling at 28 km/h in a direction 30° north of east, is 16 km due north of another ship B. The speed of ship B is 42 km/h and relative to ship A it appears to be travelling in a direction 20° north of west. Determine:

(*a*) the true direction in which ship B is travelling;
(*b*) the shortest distance that separates the two ships and the time it takes for this to occur.

19. Two men, X and Y, start walking from a cross-roads at the same time. X walks along a straight road at 7 km/h in a direction 30° west of north, while Y walks at 5 km/h along another straight road which is in a direction of 45° south of west. Determine:

(*a*) the velocity of Y relative to X;
(*b*) the distance the two men are apart after they have been walking for 30 min.

20. A ship A travelling at 18 km/h in a northerly direction observes another ship B which appears to be travelling at 25 km/h in a direction $60°$ east of south. Determine the true magnitude and direction of the velocity of ship B. If ship B was 30 km north-west of A when first observed, what would be the closest distance of approach between the two ships?

8

Kinetics—laws of motion

8.1 Kinetics

Kinetics is concerned with the motion of bodies resulting from the application of forces. Thus, we shall deal with some of the fundamental laws of engineering mechanics, namely those dealing with force and motion.

8.2 Force and motion

Motion of a body can only occur if a force is applied to it and even then the force must be sufficient to overcome any frictional resistances that may be present (ref. Chapter 5). Sir Isaac Newton, a great English scientist, put forward three laws of motion which are known as Newton's laws of Motion.

8.3 Momentum

The linear momentum of a body is defined by

$$\text{Momentum} = \text{mass} \times \text{velocity}$$
$$= mv$$

The basic unit of momentum is kg m/s.

Momentum, being a product of a scalar quantity and a vector quantity, is a vector quantity and can therefore be treated in the same manner as other vector quantities which we have used, i.e. it can be represented by a line drawn to a suitable scale and can also be resolved into components.

8.4 Newton's laws of motion

1. A body continues in its state of rest, or of uniform motion in a straight line unless it is acted upon by a resultant external force.
2. The rate of change of momentum of a body is proportional to the resultant force applied and takes place in the direction in which the force is applied.
3. To every acting force there is an equal and opposite reacting force.

The first law simply states that a body will not change its direction of motion, or its acceleration, unless it is compelled to do so by a force. Thus, we can define a force as that which changes, or tends to change, the motion of the body on which it acts.

The second law provides the basic kinetic equation (Section 8.5).

The third law, fundamental to the study of mechanics, is only the application to moving bodies of a law which we have already applied to bodies at rest, i.e. when a mass is suspended from a beam a downward force is exerted on the beam while the beam exerts an upward force on the mass. Thus, if a body A exerts a force on body B then body B exerts a similar force on body A. This applies for bodies which are in motion as well as those which are static.

8.5 The kinetic equation of motion

From Newton's Second Law:

Applied force $\propto$ rate change of momentum

$$\propto \frac{\text{change in momentum}}{\text{time}}$$

Thus, if a body of mass m experiences a change in velocity from v_1 to v_2 in a time t as a consequence of a resultant applied force F, then

$$F \propto \frac{mv_2 - mv_1}{t}$$

$$\propto \frac{m(v_2 - v_1)}{t}$$

$\propto ma$ from equation (7.1)

$\therefore F = ma \times \text{constant}$

Kinetics—laws of motion 137

Now, the unit of force in SI units is the newton which is the force required to give a mass of 1 kilogramme an acceleration of 1 metre/second².

$$1 \text{ [N]} = 1 \text{ [kg]} \times 1 \text{ [m/s}^2\text{]} \times \text{constant}$$

The constant therefore has a numerical value of unity, and

$$F = ma \qquad (8.1)$$

Example 8.1

A vehicle of mass 5 Mg accelerates from 10 km/h to 50 km/h in 8 s. Determine the force producing acceleration.

Solution

The acceleration of the vehicle can be obtained from equation (7.1).

$$v_2 = v_1 + at$$

where $\quad v_2 = 50 \text{ km/h} = \dfrac{50 \times 1000}{60 \times 60} = 13.9 \text{ m/s}$

$$v_1 = 10 \text{ km/h} = 2.78 \text{ m/s}; \qquad t = 8 \text{ s}$$

Then $\qquad 13.9 = 2.78 + 8a$

$$a = 1.39 \text{ m/s}^2$$

Applying equation (8.1):

$$F = ma$$
$$= 5 \times 10^3 \text{ [kg]} \times 1.39 \text{ [m/s}^2\text{]}$$
$$= 6.95 \times 10^3 \text{ N}$$
$$= 6.95 \text{ kN}$$

The force producing the acceleration of the vehicle is 6.95 kN.

Example 8.2

A car of mass 800 kg is accelerated uniformly from rest to a speed of 65 km/h in 6 s on a level road. The total resistances to motion, which can be assumed constant, are equivalent to a coefficient of friction between the tyres and the road of 0.035. Calculate:

(a) the total propulsive force created by the car;
(b) the change in momentum of the car.

Mechanical engineering science

Solution

(a) We can find the acceleration of the car from

$$v_2 = v_1 + at$$

where

$$v_2 = \frac{65 \times 1000}{60 \times 60} = 18.05 \text{ m/s}; \quad v_1 = 0; \quad t = 6 \text{ s}$$

$$18.05 = 0 + 6a$$

$$a = 3.01 \text{ m/s}^2$$

Gravitational force acting on car,

$$W = 800 \times 9.81 = 7848 \text{ N}$$

Normal reaction from road,

$$N = W = 7848 \text{ N}$$

From the relationship

$$\mu = \frac{F}{N}$$

Resistance to motion,

$$F = \mu N$$
$$= 0.035 \times 7848$$
$$= 274.5 \text{ N}$$

Let the total propulsive force be P [newtons]. Then referring to Fig. 8.1.

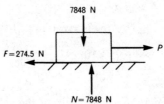

8.1 Example 8.2

Accelerating force on car $= (P - 274.5)$ N

Applying equation (8.1):

$$F = ma$$
$$P - 274.5 = 800 \times 3.01$$
$$= 2408$$
$$\therefore P = 2683 \text{ N}$$

Kinetics—laws of motion   139

The total propulsive force of the car is 2683 N.

(b) Change in momentum $= m(v_2 - v_1)$

$$= 800(18.05 - 0)$$
$$= 14\,440 \text{ kg m/s}$$

Example 8.3

A small wooden case having a mass of 16 kg rests on a horizontal surface. The case is drawn along the surface by means of a mass suspended from one end of an inextensible string, which passes over a light frictionless pulley such that the force applied to the case is horizontal. If the coefficient of friction between the case and the surface is 0.22, calculate:

(a) the minimum mass that must be suspended from the string to start the block moving;
(b) the acceleration of the case and the tension in the string when a mass of 6 kg is applied at the end of the string.

Solution

(a) In order to start the case moving it is necessary to overcome the frictional resistance to motion.

Gravitational force on case $W = 16 \times 9.81$ N $(= 157$ N$)$

Frictional resistance to motion $= \mu W$
$$= 0.22 \times 16 \times 9.81 \text{ N} = (34.5 \text{ N})$$

This is the force required to cause motion of the case.

$$\therefore \text{ Mass which must be suspended from string} = \frac{0.22 \times 16 \times 9.81}{9.81} \frac{[\text{N}]}{[\text{m/s}^2]}$$

$$= 0.22 \times 16 \text{ kg}$$
$$= 3.52 \text{ kg}$$

It will now be apparent why the gravitational force on the case was not calculated—because the 9.81 finally cancelled out.

(b) Referring to Fig. 8.2 let the tension in the string be T [newtons] and the acceleration of the system be a [metres/second^2].
For the motion of the case:

Force producing acceleration $= (T - 34.5)$ N

Applying $\qquad F = ma$

140   Mechanical engineering science

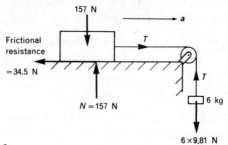

8.2 Example 8.3

gives
$$T - 34.5 = 16a \tag{1}$$

For 6-kg mass:

Gravitational force = $6 \times 9.81 = 58.86$ N

Thus,   Force producing acceleration = $(58.86 - T)$ N

$$\therefore 58.86 - T = 6a \tag{2}$$

Adding (1) and (2)

$$24.36 = 22a$$

$$a = 1.11 \text{ m/s}^2$$

In (1) $T - 34.5 = 16 \times 1.11$

$$T = 52.26 \text{ N}$$

The acceleration is 1.11 m/s² and the tension in the string is 52.26 N.

Example 8.4

Two masses, each 4 kg, are suspended from the ends of a cord passing over a light frictionless pulley. A mass of 0.5 kg is added to one of the masses. Determine:

(a) the resulting acceleration of the system,
(b) the tension in the cord.

Solution

A diagram is given in Fig. 8.3.

When the additional mass is added to one side equilibrium will be destroyed and the 4.5-kg mass will accelerate downwards while the 4-kg mass will have an equal acceleration upwards.

Kinetics—laws of motion 141

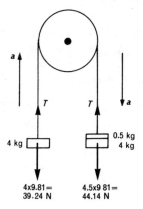

8.3 Example 8.4

Let the tension in the cord be T [newtons] and the acceleration of the system be a [metres/second^2].

METHOD 1

For 4.5-kg mass:

$$\text{Gravitational force on mass} = 4.5 \times 9.81$$
$$= 44.14 \text{ N}$$
$$\therefore \text{Force producing acceleration} = (44.14 - T) \text{ N}$$

Applying
$$F = ma$$
$$(44.14 - T) = 4.5a \qquad (1)$$

For 4-kg mass:

$$\text{Gravitational force on mass} = 4 \times 9.81$$
$$= 39.24 \text{ N}$$
$$\therefore \text{Force producing acceleration} = (T - 39.24) \text{ N}$$

Thus,
$$T - 39.24 = 4a \qquad (2)$$

Adding (1) and (2)

$$4.9 = 8.5a$$
$$a = 0.577 \text{ m/s}^2$$

From equation (2)
$$T = 39.24 + 4 \times 0.577$$
$$= 41.6 \text{ N}$$

The acceleration is 0.577 m/s^2 and the tension in the cord is 41.6 N.

METHOD 2

The mass which causes the system to accelerate is 0.5 kg. Thus,

$$\text{Accelerating force} = 0.5 \times 9.81 = 4.9 \text{ N}$$

The total mass which is accelerating $= 4 + 4.5 = 8.5$ kg

Then from
$$F = ma$$
$$4.9 = 8.5a$$
$$a = 0.577 \text{ m/s}^2$$

The tension in the cord is then obtained by considering the motion of one of the masses, as in Method 1.

The difference between the two methods should be noted. Method 1 deals with the motion of two separate masses which move as a system, whereas Method 2 considers the motion of the complete system.

Example 8.5

A railway wagon is shunted by a goods locomotive over the top of an incline of 1 in 150 at a speed of 8 km/h. The wagon rolls down the incline, which is 300 m long, and then continues along a level track until it hits a group of stationary wagons at a distance of 75 m from the bottom of the incline.

The resistance to motion can be assumed constant at 0.05 N/kg. Determine:

(*a*) the speed of the wagon at the bottom of the incline;
(*b*) the speed of the wagon when it collides with the stationary wagons.

Solution

A diagram is given in Fig. 8.4.

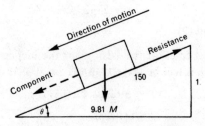

8.4 Example 8.5

(a) Let the mass of the wagon be M [kg]. Then,

Gravitational force on wagon = $9.81M$ newtons

Component of this force down the incline = $9.81M \sin \theta$

$$= \frac{9.81M}{150}$$
$$= 0.0654M \text{ newtons}$$

Resistance to motion = $0.05M$ newtons

Thus, the force producing acceleration down the incline

$$= 0.0654M - 0.05M$$
$$= 0.0154M \text{ newtons}$$

Therefore, from
$$F = ma$$
$$0.0154M = Ma$$
$$a = 0.0154 \text{ m/s}^2$$

Applying
$$v_2^2 = v_1^2 + 2as \text{ to the motion down the slope}$$

where
v_2 = speed at bottom of incline

$v_1 = 8$ km/h $= \dfrac{8 \times 1000}{60 \times 60} = 2.22$ m/s

$a = 0.0154$ m/s²

$s = 300$ m

then
$$v_2^2 = 2.22^2 + 2 \times 0.0154 \times 300$$
$$= 4.94 + 9.24$$
$$v_2 = \sqrt{14.18} = 3.76 \text{ m/s}$$
$$= \frac{3.76 \times 3600}{1000} = 13.52 \text{ km/h}$$

(b) On the level track at the bottom of the incline the only force affecting the motion will be the resistance, i.e. $0.05M$. Again, using

$$F = ma$$
$$-0.05M = Ma$$
$$a = -0.05 \text{ m/s}^2$$

144 Mechanical engineering science

(The negative sign is because the force opposes the prevailing motion.) The speed after travelling 75 m on the level track can then be obtained from

$$v_2{}^2 = v_1{}^2 + 2as$$

where v_2 = speed on collision with stationary wagons;
v_1 = 3.76 m/s; a = -0.05 m/s^2; s = 75 m

$$v_2{}^2 = 3.76^2 + 2(-0.05) \times 75$$
$$= 14.12 - 7.5$$
$$v_2 = \sqrt{6.62} = 2.57 \text{ m/s}$$
$$= 2.57 \times 3.6 = 9.25 \text{ km/h}$$

The speed of the wagon at the bottom of the incline will be 13.52 km/h, and will be 9.25 km/h upon impact with the stationary wagons.

8.6 Conservation of momentum

From Newton's third law of motion another important principle, known as the Principle of the Conservation of Linear Momentum, is obtained. This states that: The total momentum of a system, in a certain direction, remains constant unless an external force is applied to the system in that direction.

This principle enables impact problems, where only internal forces are involved, to be easily solved.

Consider the firing of a gun. The equal and opposite forces exerted by the gun on the projectile and the projectile on the gun will cause motion of each, but as there are no external forces involved it follows that the change in momentum of the projectile due to firing must be equal and opposite to the change in momentum of the gun. The difference in the resulting velocities of the projectile and the gun is due entirely to their difference in mass.

Examples 8.6 and 8.7 illustrate the application of the Principle of the Conservation of Momentum.

Example 8.6

An object A of mass 5 kg is moving with a velocity of 18 m/s when it collides with a second object B, of mass 2 kg, which is moving on the same line, with a velocity of 4 m/s. After impact the two objects move together.

Determine their common velocity if:
(a) A and B are initially moving in the same direction;
(b) A and B are initially moving in opposite directions.

Solution

(a) Let the common velocity after impact be v_1' as shown in Fig. 8.5.

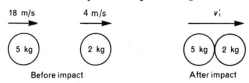

8.5 Conservation of momentum—Example 8.6

By the Principle of the Conservation of Linear Momentum:

Momentum before impact = momentum after impact
$$5 \times 18 + 2 \times 4 = (5+2)v_1'$$
$$v_1' = 14 \text{ m/s}$$

The common velocity after impact is 14 m/s in the same direction as A and B were moving before impact.

(b) Let the common velocity after impact be v_2' as shown on Fig. 8.6.

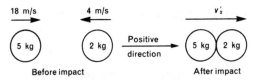

8.6 Conservation of momentum—Example 8.6

In this case, the momentum of the 2-kg mass before impact is in the opposite sense to that of the 5-kg mass. Applying

Momentum before impact = momentum after impact
$$5(+18) + 2(-4) = (5+2)v_2'$$
$$v_2' = 11.7 \text{ m/s}^2$$

The common velocity after impact is 11.7 m/s in the same direction as A was moving before impact.

Example 8.7

A pile-driver having a mass of 300 kg is used to drive a pile of mass 500 kg into the ground. If the pile-driver falls through a height of 4 m onto the pile, how far will the pile be driven into the ground? Assume that the pile and pile-driver remain together after impact and that the resisting force of the ground is 140 kN.

Solution

The problem is solved by finding the velocity of the pile-driver on impact with the pile and then applying the Principle of the Conservation of Linear Momentum to derive the common velocity of the pile and pile-driver after impact. Using this velocity, and the ground resistance, we can apply Newton's second law to calculate the deceleration of the pile. Then, by applying an appropriate equation of motion the penetration of the pile is obtained.

The velocity of the pile-driver on impact (v_2) is obtained from

$$v_2^2 = v_1^2 + 2as$$

where $\quad v_1 = 0; \quad a = 9.81 \text{ m/s}^2; \quad s = 4 \text{ m}$

$$v_2^2 = 2 \times 9.81 \times 4$$

$$v_2 = 8.86 \text{ m/s}$$

Applying the Principle of the Conservation of Linear Momentum

$$\frac{\text{Momentum of pile-driver}}{\text{before impact}} = \frac{\text{momentum of pile and}}{\text{pile-driver after impact}}$$

$$300 \times 8.86 = (300+500)V$$

where V = common velocity of pile and pile-driver after impact. Thus

$$V = \frac{300 \times 8.86}{800}$$

$$= 3.33 \text{ m/s}$$

If the acceleration of the pile and pile-driver after impact is a, then from

$$F = ma$$

we obtain $\quad -140 \times 10^3 \text{ [N]} = 800 \text{ [kg]} \times a$

$$a = -175 \text{ m/s}^2$$

Then, applying $\qquad v_2^2 = v_1^2 + 2as$

Kinetics—laws of motion   147

to the motion after impact, where

v_2 = final velocity = 0; v_1 = initial velocity = 3.33 m/s
$a = -175$ m/s^2; s = distance moved

gives
$$0^2 = 3.33^2 + 2(-175)s$$
$$s = 0.0317 \text{ m} = 31.7 \text{ mm}$$

The pile will be driven 31.7 mm into the ground.

8.7 Impact of a fluid jet

The impact of a fluid jet upon a fixed vane is a good example of the application of the three laws of motion. If the jet is deflected from its line of motion then, from the first law, a force must be acting upon it. The second law of motion tells us that the force which is acting must be proportional to the rate of change of momentum and act in the direction in which the change in momentum occurs. Finally, the third law says that the force exerted by the vane on the jet to change its momentum must be equal and opposite to the force exerted by the jet on the vane.

When dealing with problems on the impact of a fluid jet we are usually concerned with the jet flow rate. Thus, the second law of motion can be written in the form:

$$\text{Force} = \text{rate of change of momentum}$$

$$= \frac{\text{mass} \times \text{velocity change}}{\text{time taken}}$$

$$= \text{mass flow rate} \times \frac{\text{change of velocity in the}}{\text{direction of the force}}$$

Example 8.8
A water jet of 60 mm diameter impacts perpendicularly upon a fixed plate with a velocity of 40 m/s. Calculate the force acting on the plate.

Solution
When the jet of water strikes the plate it will spray out radially, i.e. it is deflected through an angle of 90°. Thus, the jet will lose all its momentum at right angles to the plate, as illustrated in Fig. 8.7.

148   Mechanical engineering science

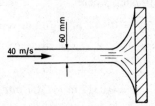

8.7 Impact of a fluid jet—Example 8.8

$$\text{Area of jet} = \frac{\pi}{4} \times \left(\frac{60}{1000}\right)^2 \text{ m}^2 = \frac{9\pi}{10^4} \text{ m}^2$$

Volume flow rate = area × velocity

$$= \frac{9\pi}{10^4} \text{ [m}^2\text{]} \times 40 \text{ [m/s]}$$

$$= 0.036\pi \text{ m}^3\text{/s}$$

Mass flow rate = volume flow rate × density

$$= 0.036\pi \text{ [m}^3\text{/s]} \times 10^3 \text{ [kg/m}^3\text{]}$$

$$= 36\pi \text{ kg/s}$$

Then, Force on plate = mass flow rate × change in velocity

$$= 36\pi \text{ [kg/s]} \times 40 \text{ [m/s]}$$

$$= 4520 \text{ N}$$

$$= 4.52 \text{ kN}$$

The force on the plate is 4.52 kN.

Problems

1. A lorry of mass 6.5 Mg is accelerated uniformly from rest to a speed of 25 km/h on a level road in 20 s. Assuming that the resistance to motion remains constant at 0.2 N/kg determine the propelling force acting on the lorry.

2. A road vehicle has a mass of 1600 kg and is uniformly accelerated from rest to a speed of 60 km/h in a distance of 48 m on a level surface against a tractive resistance of 0.18 N/kg. Determine:

(a) the propulsive force of the vehicle;
(b) the change in momentum.

3. A train having a mass of 300 Mg accelerates uniformly from 20 km/h to 120 km/h in a distance of 3.2 km. Determine:

Kinetics—laws of motion 149

(a) the rate of acceleration;
(b) the force producing the acceleration;
(c) the change in the momentum of the train.

4. A body having a mass of 90 kg is drawn up a plane inclined at 12° to the horizontal by a force of 504 N acting parallel to the plane. If the coefficient of friction between the body and the plane is 0.3, calculate the acceleration of the body up the incline.

5. A mass of 50 kg is drawn up a rough plane inclined at 16° to the horizontal by a force applied parallel to the plane. Assuming a coefficient of friction of 0.33 calculate the force required to produce an acceleration of 0.15 m/s^2.

6. A body having a mass of 14 kg is pulled up a rough plane inclined at 18° to the horizontal by a force of 88 N applied parallel to the plane. Assuming the coefficient of friction between the body and the plane is 0.28, determine:

(a) the acceleration of the body up the plane;
(b) the time to travel 50 m up the plane;
(c) the work done in overcoming friction during the first 15 s and second 15 s respectively.

7. A train of mass 400 Mg is hauled up an incline 1.6 km long and having a slope of 1 in 120. The frictional resistance, which can be assumed constant, is 0.062 N/kg. The diesel power unit exerts a constant tractive force and the speed of the train is 64 km/h at the bottom of the incline and 40 km/h at the top. Determine:

(a) the tractive effort exerted by the diesel unit;
(b) the time taken to climb the incline.

8. A body of mass 30 kg rests on a rough plane which is inclined at 25° to the horizontal. What force, applied parallel to the plane, is required to move the body up the plane (a) at a steady speed, (b) with an acceleration of 0.4 m/s^2?

Assume the coefficient of friction is 0.22.

9. A truck of mass 1100 kg is being hauled up a plane, inclined at 20° to the horizontal, by means of a haulage rope which is parallel to the plane. If the resistance to motion is equivalent to a force of 1750 N and the constant pull exerted by the haulage rope is 5800 N determine the acceleration of the truck up the incline.

How long would it take for the truck to be hauled through a distance of 360 m up the incline, starting from rest.

10. A train having a mass of 360 Mg is travelling along a level track at a speed of 110 km/h. It then climbs an incline of 1 in 160, which is 900 m long, with the tractive effort remaining the same as on the level track. Assuming that the frictional and other resistances to motion remain constant at 60 N/Mg calculate:

(a) the retardation of the train while climbing the incline;
(b) the speed at the top of the incline.

11. A body having a mass of 70 kg is pulled across a horizontal floor by a horizontal force of 260 N. The body travels from rest through a distance of 60 m in a time of 10 s under the influence of the force. If the force is removed at the end of this time what further distance will the body travel before coming to rest?

12. A railway wagon of mass 5000 kg rests on an incline of 1 in 120 with the brake applied. The brake is now released. If the resistances to motion are equivalent to a coefficient of friction of 0.005, determine:

(a) the force which would be required to stop the truck rolling down the incline;
(b) the acceleration of the truck down the incline;
(c) the work done against friction when the truck rolls through a distance of 150 m down the incline.

13. A casting of mass 2250 kg is raised vertically by means of a hoisting gear which is driven by an electric motor. If the casting is raised with a uniform acceleration such that after a time of 3 s it is at a height of 5 m determine:

(a) the force in the steel lifting wire;
(b) the safety factor of the lifting wire if it has an effective diameter of 18 mm and the yield strength of the material is 285 MN/m^2.

14. A packing case having a mass of 800 kg is lifted with a uniform acceleration by a crane such that after being raised through a height of 12 m it has a velocity of 9 m/s. Calculate the force in the lifting cable.

15. A train of mass 350 Mg is hauled by a locomotive having a mass of 100 Mg. The train starts from a station and travels along a level track for a distance of 1.5 km, by which time its speed is 56 km/h. It then starts to climb an incline of 1 in 80, 1.2 km long, and the tractive effort of the locomotive, which remained constant while on the level, is increased by 15%.

Assuming the resistance to motion to be 0.058 N/kg for both the locomotive and the train, determine:

Kinetics—laws of motion 151

(*a*) the tractive effort exerted by the locomotive on the level track;
(*b*) the speed of the train at the top of the incline;
(*c*) the time to reach the top of the incline from rest.

16. An express train having a mass of 480 Mg is hauled by an electric traction unit. On leaving a station and travelling on a level track the train accelerates uniformly to 90 km/h in a distance of 1.4 km. If the frictional and other resistances amount to 60 N/Mg determine the required diameter of the drawbar between the traction unit and the train if there is to be a safety factor of 15 based upon a yield stress of 275 MN/m².

17. Two masses, of 5 kg and 8 kg respectively, are suspended from the ends of a cord which passes over a light frictionless pulley. When the system is released, calculate:

(*a*) the acceleration of the masses;
(*b*) the tension in the cord.

18. Two masses, each of 3 kg, are suspended from the ends of a wire passing over a light frictionless pulley. If an additional mass of 0.4 kg is now added to one of the masses, determine:

(*a*) the acceleration of the system;
(*b*) the tension in the wire;
(*c*) the time required for the masses to move through a height of 2 m.

19. A body of mass 4.5 kg rests on a table, the coefficient of friction between the two being 0.12. This body is connected to another body by means of an inextensible string which passes over a smooth pulley and hangs vertically downwards. The mass of the second body is 0.9 kg. Calculate:

(*a*) the tension in the string;
(*b*) the acceleration of the masses;
(*c*) the velocity of the masses 2 s after starting from rest.

20. A string passes over a light frictionless pulley and carries a mass of 4 kg at each of its ends. An additional mass is added to one of the masses such that when released the system accelerates uniformly so that the masses move through a distance of 1.514 m in a time of 5 s. Determine the additional mass added and the tension in the cord.

21. A cord passes over a smooth frictionless pulley and carries a mass of 6 kg at each of its ends. What mass must now be added to one of the masses to produce an acceleration of 0.8 m/s²? What is the tension in the cord under these conditions?

152   Mechanical engineering science

22. A railway wagon having a mass of 6 Mg is travelling at 10 km/h along a straight level track when it collides with another truck having a mass of 2.5 Mg which is moving in the same direction with a velocity of 2 km/h. After impact the two trucks move together. Calculate:
(a) the common velocity after impact;
(b) the distance that the wagons move together before coming to rest if the resistance to motion is 0.066 N/kg.

23. A ball, of mass 0.9 kg, is moving on a level surface at 16 km/h when it hits another ball, of mass 0.5 kg, which is moving in the opposite direction along the same line with a velocity of 4 km/h. If the two balls travel together after impact what will be their common velocity?

24. A truck having a mass of 8000 kg is travelling at 4 m/s on a level track when it collides with a second truck having a mass of 6000 kg which is travelling on the same line, but in the opposite direction, at a speed of 6 m/s. Assuming that the two trucks remain together after impact what will be the magnitude and direction of their common velocity?

25. A pile driver hammer of mass 700 kg has a vertical velocity of 24 m/s when it strikes an inelastic pile of mass 360 kg. If, after impact, the driver and the pile move together and the pile is driven 0.10 m into the ground determine:
(a) the common velocity of the hammer and pile after impact;
(b) the resistance exerted by the ground.

26. A body having a mass of 200 kg is used to drive a steel pile of mass 900 kg into the ground. If the body is dropped through a height of 3.5 m before it hits the pile what should be the penetration of the pile if the ground resisting force is 80 kN and there is no rebound of the body after impact?

27. A pile-driver of 500 kg mass is to be used to drive a pile of 900 kg mass into the ground. If the ground resistance is 60 kN and it is required to drive the pile 0.12 m into the ground by means of a single blow, from what height should the pile driver be allowed to fall?

28. A jet of water of 75 mm diameter impacts perpendicularly upon a fixed flat plate with a velocity of 25 m/s. Calculate the force acting upon the plate.
Assume the density of water to be 1 Mg/m^3.

29. When a jet of water of 50 mm diameter impinges on a flat plate the force on the plate is 1.77 kN. Determine the velocity of the jet.

9

Angular motion

9.1 Angular motion

When a rigid body, such as a wheel, rotates about a fixed axis all points on the wheel are constrained to move in a circular path. It follows that in any given period of time all points on the wheel will complete the same number of revolutions about the axis of rotation and so we talk of the speed of rotation in revolutions/minute (rev/min.). However, it is frequently more convenient to express the speed of rotation in terms of the angle turned through in unit time.

9.2 The radian

The radian, which is the unit of angular measure, is the angle subtended at the centre of a circle by an arc equal to the radius. This is shown in Fig. 9.1 where the angle AOB is equal to one radian.

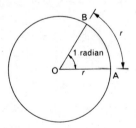

9.1 Angular displacement

As the circumference of a circle is equal to $2\pi r$ it follows that there are 2π radians in a circle. Thus

2π radian $= 360°$

1 radian $= 57.3°$

9.3 Angular velocity

Consider a rigid body rotating about a fixed axis at a uniform speed of rotation of n rev/s. As the body turns through 2π radians (rad) in each revolution the angular velocity ω (omega) is given by

$$\omega = 2\pi n \text{ rad/s} \qquad (9.1)$$

Also, if the body rotates through an angle of θ radians in a time of t seconds then, providing that the motion is uniform

$$\text{Angular velocity } \omega = \frac{\theta}{t} \text{ rad/s} \qquad (9.2)$$

Note that angular velocity has no linear dimensions.

9.4 Relationship between angular and linear velocity

Consider a wheel which is rotating about an axis through O at a uniform speed. Suppose that in a period of time t a point on the wheel, at a radius r from the axis of rotation, moves from P_1 to P_2 such that the angle turned through is θ, as shown in Fig. 9.2.

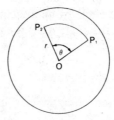

9.2 Angular motion

Then, from our definition of speed;

$$\text{Speed} = \frac{\text{distance covered}}{\text{time taken}}$$

$$v = \frac{P_1 P_2}{t}$$

Angular motion   155

$$v = \frac{r\theta}{t}$$

From equation (9.2) $\theta/t = \omega$

$$\therefore v = r\omega \qquad (9.3)$$

Putting this equation into words:

Linear speed = radius from axis × angular velocity

9.5 Constant angular acceleration

Suppose the angular velocity of a body increases uniformly from ω_1 to ω_2 in a period of time t. The angular acceleration α (alpha) is obtained from:

$$\text{Angular acceleration} = \frac{\text{change in angular velocity [rad/s]}}{\text{time taken [s]}}$$

$$\alpha = \frac{(\omega_2 - \omega_1)}{t} \text{ rad/s}^2$$

$$\omega_2 = \omega_1 + \alpha t \qquad (9.4)$$

Angular displacement = average angular velocity × time

$$\theta = \frac{(\omega_1 + \omega_2)}{2} \times t \qquad (9.5)$$

Substituting for ω_2 from equation (9.4) gives:

$$\theta = \frac{(\omega_1 + \omega_1 + \alpha t)}{2} t$$

$$\theta = \omega_1 t + \tfrac{1}{2}\alpha t^2 \qquad (9.6)$$

Or, substituting for ω_1 from equation (9.4) into (9.5):

$$\theta = \omega_2 t - \tfrac{1}{2}\alpha t^2 \qquad (9.7)$$

Squaring equation (9.4)

$$\omega_2{}^2 = (\omega_1 + \alpha t)^2$$
$$= \omega_1{}^2 + 2\alpha t \omega_1 + \alpha^2 t^2$$
$$= \omega_1{}^2 + 2\alpha(\omega_1 t + \tfrac{1}{2}\alpha t^2)$$
$$= \omega_1{}^2 + 2\alpha\theta \qquad (9.8)$$

156   Mechanical engineering science

The equations (9.4) to (9.8) are the equations of angular motion for uniform angular acceleration. They are summarised below and should be compared with the equations governing linear motion with uniform acceleration given in Section 7.8 (page 118).

$$\omega_2 = \omega_1 + \alpha t \qquad \omega_1 = \text{initial angular velocity}$$

$$\theta = \frac{(\omega_1 + \omega_2)}{2} \times t \qquad \omega_2 = \text{final angular velocity}$$

$$\theta = \omega_1 t + \tfrac{1}{2}\alpha t^2 \qquad \alpha = \text{angular acceleration}$$

$$\omega_2^2 = \omega_1^2 + 2\alpha\theta \qquad \theta = \text{angle rotated through}$$

$$\theta = \omega_2 t - \tfrac{1}{2}\alpha t^2 \qquad t = \text{time taken}$$

Example 9.1

A flywheel having a diameter of 2 m is accelerated uniformly from rest so that after 1 min its speed of rotation is 240 rev/min. After running at this speed for a further 5 min it is brought to rest with a uniform retardation in a time of 45 s. Determine:

(a) the angular acceleration of the flywheel;
(b) the total number of revolutions made by the flywheel;
(c) the maximum speed of a point on the circumference of the flywheel.

Solution

There are three separate periods in the motion of the flywheel. These are:

(1) acceleration to 240 rev/min in 1 min;
(2) constant speed of 240 rev/min for 5 min;
(3) stopping of wheel with uniform retardation in 45 s.

(a) The angular acceleration is obtained from the acceleration stage by applying equation (9.4):

$$\omega_2 = \omega_1 + \alpha t$$

where

$$\omega_1 = 0; \quad \omega_2 = 240 \text{ rev/min} = \frac{240 \times 2\pi}{60} = 8\pi \text{ rad/s}$$

$$t = 60 \text{ s}; \quad \alpha = \text{angular acceleration}$$

Then
$$8\pi = 0 + 60\alpha$$
$$= 0.418 \text{ rad/s}^2$$

The acceleration of the flywheel is 0.418 rad/s²

(b) To determine the total number of revolutions of the flywheel each period of the motion must be considered separately.

Let the angle turned through during the acceleration period be θ_1.
Applying equation (9.5)
$$\theta_1 = \left(\frac{\omega_1 + \omega_2}{2}\right) \times t$$
$$= \left(\frac{0 + 8\pi}{2}\right) \times 60$$
$$= 240\pi \text{ rad}$$

During the constant speed stage
$$\theta_2 = \omega_2 t_2$$
where $\quad \omega_2 = $ constant angular velocity $= 8\pi$ rad/s
and $\quad t_2 = 5$ min $= 300$ s
$$\theta_2 = 8\pi \times 300$$
$$= 2400\pi \text{ rad}$$

If θ_3 is the angle rotated through during the retardation period. Then,
$$\theta_3 = \left(\frac{\omega_2 + \omega_3}{2}\right) \times t_3$$
where $\quad \omega_2 = 8\pi$ rad/s; $\quad \omega_3 = 0;$ $\quad t_3 = 45$ s
$$\theta_3 = \left(\frac{8\pi + 0}{2}\right) \times 45$$
$$= 180\pi \text{ rad}$$

Thus, the total angle turned through by the flywheel, θ, is
$$\theta = \theta_1 + \theta_2 + \theta_3$$
$$= 240\pi + 2400\pi + 180\pi$$
$$= 2820\pi \text{ rad}$$

But there are 2π radians in one revolution.

$\therefore$ Number of revolutions $= \dfrac{2820\pi}{2\pi} = 1410$

(c) The maximum rim speed v is obtained from equation (9.3).

$$v = r\omega$$

where $r = 1$ m and $\omega = $ max. angular velocity $= 8\pi$ rad/s

$$\therefore v = 1 \times 8\pi$$
$$= 25.1 \text{ m/s}$$

The maximum speed of a point on the rim of the flywheel is 25.1 m/s.

Example 9.2

A wheel is rotating at 2500 rev/min when it is subjected to a braking torque which produces an angular retardation of 12 rad/s². Determine:
(a) the time taken to slow the wheel to 1000 rev/min;
(b) the angular velocity of the wheel after it has rotated through 200 revolutions.

Solution

Applying $\quad \omega_2 = \omega_1 + \alpha t \quad$ (equation (9.4))

where $\quad \omega_1 = $ initial angular velocity $= \dfrac{2500 \times 2\pi}{60}$ rad/s

$\omega_2 = $ angular velocity after time $t = \dfrac{1000 \times 2\pi}{60}$ rad/s

$\alpha = $ angular acceleration $= -12$ rad/s²

Then $\quad \dfrac{1000 \times 2\pi}{60} = \dfrac{2500 \times 2\pi}{60} + (-12)t$

$$12t = \dfrac{1500 \times 2\pi}{60}$$

$$t = 13.1 \text{ s}$$

(b) Applying $\quad \omega_2{}^2 = \omega_1{}^2 + 2\alpha\theta \quad$ (equation (9.8))

where $\quad \omega_2 = $ angular velocity after rotating through 200 rev

$$\omega_1 = \dfrac{2500 \times 2\pi}{60} = 262 \text{ rad/s}$$

$$\alpha = -12 \text{ rad/s}^2; \ \theta = 200 \times 2\pi \text{ rad}$$

gives
$$\omega_2{}^2 = 262^2 + 2 \times (-12) \times 400\pi$$
$$= 68\,644 - 30\,160$$
$$= 38\,484$$
$$\omega_2 = 196 \text{ rad/s}$$
$$= \frac{196}{2\pi} \times 60 = 1872 \text{ rev/min}$$

9.6 Centripetal acceleration

Consider a body of mass m to be moving on a circular path of radius r, as shown in Fig. 9.3(i), with a constant angular velocity ω.

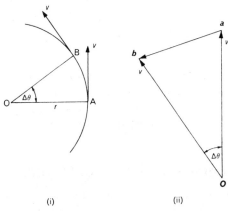

(i) (ii)

9.3 Centripetal acceleration

When the body is at A it possesses an instantaneous linear velocity, v, tangential to the circle. Suppose that during a small period of time Δt, the body moves from A to B, turning through the small angle $\Delta\theta$. Now although the body is moving at a constant speed it is not moving along the same line of action and therefore there is a change in velocity. Thus, the body is subject to an acceleration in the same direction as the change in velocity.

Referring to the velocity vector diagram (Fig. 9.3(ii)), vector **Oa** represents the velocity at A while vector **Ob** represents the velocity at B. It will be seen that there is a vector change in velocity during the motion from A to B of magnitude **ab**, acting in a radial direction.

$$\text{Vector change in velocity} = \mathbf{ab}$$
$$= v \cdot \Delta\theta$$

160 Mechanical engineering science

But from equation (9.2)

$$\theta = \omega t$$

$$\therefore \Delta\theta = \omega \cdot \Delta t$$

$$= \frac{v}{r} \cdot \Delta t \quad \text{(from equation (9.3))}$$

Thus, Vector change in velocity $= \dfrac{v^2}{r} \cdot \Delta t$

But, Acceleration $= \dfrac{\text{change in velocity}}{\text{time taken}}$

$$= \frac{v^2}{r} \cdot \frac{\Delta t}{\Delta t}$$

$$= \frac{v^2}{r} = \omega^2 r \qquad (9.9)$$

This radial acceleration is known as centripetal acceleration and for a uniform speed of rotation the centripetal acceleration is constant. Any angular acceleration to which the body is subjected will be in addition to the centripetal acceleration present.

9.7 Centripetal and centrifugal forces

We already know that a body cannot be subjected to an acceleration unless a force is acting to cause that acceleration. Thus, to produce a centripetal acceleration a centripetal force must be acting radially inwards on the body. From Newton's Second Law:

Centripetal force $=$ mass $\times$ centripetal acceleration

$$= m\frac{v^2}{r} = m\omega^2 r \qquad (9.10)$$

This force, which must be applied externally to a body, can be created in various ways. In the case of a mass on the end of a string or carried on a rotating arm the tension in the string or in the arm provides the force. With a train, however, this force is provided by the side thrust of the rails on the flanges of the wheels. In other cases the centripetal force may arise from frictional effects. This is the case when a car negotiates a bend and if the road conditions are such that insufficient frictional force can be created at the tyres then a skid will result.

Angular motion

By Newton's third law there must be an opposing reaction force of equal magnitude to the centripetal force. This force therefore acts radially outwards and is known as the centrifugal force. It must be appreciated that the centrifugal force is NOT applied to the body, e.g. in the case of a train on a curve, the train pushes outwards on the rails (centrifugal) and the rails exert a force of equal magnitude on the train (centripetal).

Example 9.3

A steel ball having a mass of 1 kg is rotated at a speed of 140 rev/min in a horizontal circular motion of radius 0.6 m. The ball is connected to the centre of rotation by a steel wire of diameter 0.6 mm. Calculate:
(a) the centripetal acceleration;
(b) the tension in the wire;
(c) the speed of rotation at which the wire would fracture if its tensile strength is 540 MN/m².

Would the tension in the wire be any different if the motion were not in a horizontal plane?

Solution

(a) From equation (9.9)

$$\text{Centripetal acceleration} = \omega^2 r$$

where $\omega = \dfrac{140 \times 2\pi}{60} = 14.65 \text{ rad/s}; \quad r = 0.6 \text{ m}$

Then,

$$\text{Centripetal acceleration} = 14.65^2 \, [\text{rad}^2/\text{s}^2] \times 0.6 \, [\text{m}]$$
$$= 128.8 \text{ m/s}^2$$

(b) Tension in wire = centripetal force
$$= m\omega^2 r \quad \text{(equation (9.10))}$$
$$= 1 \, [\text{kg}] \times 128.8 \, [\text{m/s}^2]$$
$$= 128.8 \text{ N}$$

(c) Stress in wire $= \dfrac{\text{force}}{\text{area}}$

∴ Required force to fracture wire $= \text{tensile strength} \times \text{area}$

$$= 540 \times 10^6 \, [\text{N/m}^2] \times \frac{\pi}{4} (0.6 \times 10^{-3})^2 \, [\text{m}^2]$$
$$= 152.6 \text{ N}$$

Let the angular velocity at which fracture of the wire would occur be ω_1.

Then, Centripetal force $= m\omega_1^2 r$
$$= 1 \times \omega_1^2 \times 0.6$$

This force must equal the force required to break wire, i.e.
$$0.6\omega_1^2 = 152.6$$
$$\omega_1^2 = 254.3$$
$$\omega_1 = 15.95 \text{ rad/s}$$
$$= \frac{15.95}{2\pi} \times 60$$
$$= 152.4 \text{ rev/min}$$

The speed of rotation at which the wire would fracture is 152.4 rev/min.

If the motion were not in a horizontal plane then the gravitational force on the ball would affect the tension in the wire.

Consider the case of motion in a vertical plane.

Gravitational force on ball $= 9.81 \times 1 = 9.81$ N

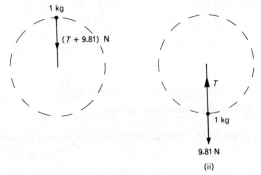

9.4 Rotation of a body in a vertical plane

When the ball is at the top of the circle (Fig. 9.4(i))
$$T + 9.81 = \text{centripetal force}$$
$$\therefore T = \text{centripetal force} - 9.81$$

While for the body at the bottom of the circle (Fig. 9.4(ii))
$$T - 9.81 = \text{centripetal force}$$
$$\therefore T = \text{centripetal force} + 9.81$$

Thus, the latter case provides the greater tension in the wire and, in fact, is the greatest value possible for any plane of rotation.

9.8 Balancing of co-planar masses

(a) STATIC BALANCE

We have seen earlier that for a body to be in equilibrium statically there must be no resultant force or resultant moment acting on the body. Thus, consider a system of three co-planar masses m_1, m_2, and m_3 at radii r_1, r_2, and r_3 from an axis O and positions θ_1, θ_2, and θ_3 relative to axis OX, as shown in Fig. 9.5. The respective gravitational forces acting on the masses will be 9.81 × mass.

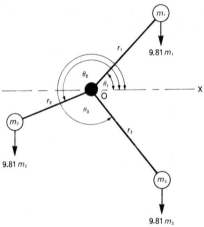

9.5 Static balance of co-planar masses

For this system to be balanced the total moment about O must be zero. Thus,

$$9.81 m_1 r_1 \cos \theta_1 + 9.81 m_2 r_2 \cos \theta_2 + 9.81 m_3 r_3 \cos \theta_3 = 0$$

or
$$m_1 r_1 \cos \theta_1 + m_2 r_2 \cos \theta_2 + m_3 r_3 \cos \theta_3 = 0 \quad (9.11)$$

This single relationship is not sufficient for the system to be balanced for all angular positions. If the system rotates through 90° so that the X direction is now vertical the total moment about O must again be zero, giving:

$$9.81 m_1 r_1 \sin \theta_1 + 9.81 m_2 r_2 \sin \theta_2 + 9.81 m_3 r_3 \sin \theta_3 = 0$$
$$m_1 r_1 \sin \theta_1 + m_2 r_2 \sin \theta_2 + m_3 r_3 \sin \theta_3 = 0 \quad (9.12)$$

Now a system which is in static balance for any two directions at right angles will also be balanced for any intermediate position. Therefore, the equations (9.11) and (9.12) enable the balance of a system to be checked and if the given system is not balanced they are sufficient to enable the magnitude and position of a suitable balancing mass to be determined.

(*b*) ROTATIONAL BALANCE

When a system of masses rotates about an axis there will be a centrifugal force exerted on the shaft by each of the masses. If the system is to rotate smoothly these centrifugal forces must be arranged so that the total resultant force on the shaft is zero. When this state of affairs is achieved the system is said to be balanced. The subject of balancing is very extensive and complex and we can only deal in this book with the balancing of co-planar masses.

Consider the system of Fig. 9.5 to be rotating about axis O at an angular speed ω. Each of the masses will exert a centrifugal force of $mr\omega^2$ on the shaft, as shown on Fig. 9.6.

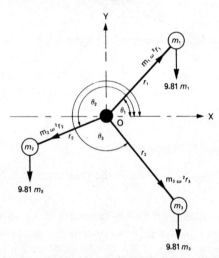

9.6 Rotational balance of co-planar masses

If these forces are in balance there must be no resultant force in the X or Y directions.

In X direction:

$$m_1 r_1 \omega^2 \cos \theta_1 + m_2 r_2 \omega^2 \cos \theta_2 + m_3 r_3 \omega^2 \cos \theta_3 = 0$$

In Y direction:

$$m_1 r_1 \omega^2 \sin \theta_1 + m_2 r_2 \omega^2 \sin \theta_2 + m_3 r_3 \omega^2 \sin \theta_3 = 0$$

As ω^2 is constant throughout, these equations reduce to forms identical to equations (9.11) and (9.12).

Thus, for a system of co-planar masses, the conditions for static and dynamic balance are identical.

Equations (9.11) and (9.12) can be represented by a vector diagram as shown in Fig. 9.7.

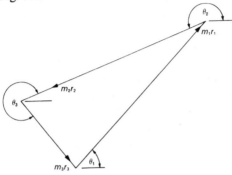

9.7 Vector diagram for co-planar balancing

If such a vector diagram does not close then the masses are not in balance. The closing line does, however, give the angular position and mr value of the mass required to balance the system. This is illustrated by Example 9.4.

Example 9.4

A rotating shaft has three arms attached to it such that they lie in a plane perpendicular to the axis of rotation. Two of the arms are inclined at 60° to each other and have masses of 1.5 kg and 3.5 kg fixed at distances of 0.15 m and 0.20 m respectively from the axis of rotation. Determine where a mass of 2.5 kg should be positioned on the third arm to balance the system.

Solution

Let the mass of 2.5 kg be positioned at a radius r_3 on the third arm, positioned at an angle θ_3 with the arm carrying the mass of 1.5 kg as shown in Fig. 9.8.

166   Mechanical engineering science

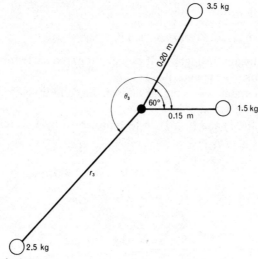

9.8 Example 9.4

The vector product of mass × radius (mr) is now calculated.
For 1.5 kg mass:

$$mr = 1.5 \times 0.15 = 0.225 \text{ kg m}$$

For 3.5 kg mass

$$mr = 3.5 \times 0.20 = 0.70 \text{ kg m}$$

Vectors **ab** and **bc** are now drawn to scale, as shown in Fig. 9.9, in directions parallel to the respective arms.

To close the diagram we require vector **ca**, and from Fig. 9.9 **ca** = 0.835 kg m acting at $\theta_3 = 226°$.

Then for the third mass, $m_3 r_3 = 0.835$ kg m.

But the third mass is specified as 2.5 kg. Thus

$$r_3 = \frac{0.835}{2.5}$$

$$= 0.334 \text{ m}$$

The mass of 2.5 kg must be positioned at a radius of 0.334 m from the centre of rotation on an arm making an angle of 226° with the arm carrying the 1.5 kg mass, the angle being measured in the same direction as the angle of 60°.

This problem can also be solved analytically by applying equations (9.11) and (9.12).

Angular motion

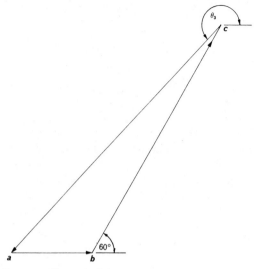

9.9 Vector diagram—Example 9.4

Thus, referring to Fig. 9.8 and applying equation (9.11) gives:

$$1.5 \times 0.15 \cos 0° + 3.5 \times 0.2 \cos 60° + 2.5 r_3 \cos \theta_3 = 0$$
$$0.225 + 0.35 + 2.5 r_3 \cos \theta_3 = 0$$
$$0.575 + 2.5 r_3 \cos \theta_3 = 0 \qquad (1)$$

and applying equation (9.12):

$$1.5 \times 0.15 \sin 0° + 3.5 \times 0.2 \sin 60° + 2.5 r_3 \sin \theta_3 = 0$$
$$0.6062 + 2.5 r_3 \sin \theta_3 = 0 \qquad (2)$$

Rewriting equations (2) and (1)

$$2.5 r_3 \sin \theta_3 = -0.6062$$

and
$$2.5 r_3 \cos \theta_3 = -0.575$$

$$\therefore \tan \theta_3 = \frac{-0.6062}{-0.575}$$
$$= +1.054$$
$$\theta_3 = 46° \, 31' \text{ or } 226° \, 31'$$

As the system could not be balanced with $\theta_3 = 46° \, 31'$ the only possible solution is $\theta_3 = 226° \, 31'$.

From (1) $r_3 = \dfrac{-0.575}{2.5 \cos 226° 31'}$

$= +0.334$ m

9.9 Relative velocity of two points on a rotating link

In Section 7.11, we dealt with the relative velocity between two bodies that were unconnected. We must now deal with the relative velocity between points on a rigid rotating body.

Consider a rigid link AB, of length l, forming part of a machine, to be moving such that, at a certain instant, A and B have velocities v_a and v_b respectively in the directions shown in Fig. 9.10.

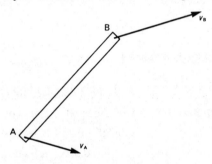

9.10 Relative velocity of points on a rotating link

As the points A and B are constrained to remain a fixed distance apart, it follows that any relative velocity between A and B cannot have a component along the line AB. Thus, any relative velocity must be at right angles to the line AB since this is the only direction which will not produce a component along AB. Drawing the velocity vector diagram by the method of Section 7.11 gives the diagram shown in Fig. 9.11.

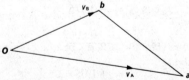

9.11 Velocity vector diagram for rotating link

The velocity of A relative to B, which is given by vector **ba**, is at right angles to AB and therefore, relative to B, A is apparently rotating with a tangential velocity **ba**. But,

Angular velocity of link = $\dfrac{\text{tangential velocity}}{\text{radius}}$

$$\therefore \omega = \frac{ab}{l}$$

If we had considered the velocity of B relative to A a similar result would have been obtained, namely that, relative to A, B is apparently rotating with an angular velocity ω. This is an obvious result since the link AB can only have one value of angular velocity.

The motion of any other point on the link, relative to either A or B, must also be at right angles to the link and its tangential velocity will depend upon its distance from the ends of the link. This is dealt with in Example 9.5.

Example 9.5

An internal combustion engine has a crank 0.15 m long and a connecting rod 0.375 m long. When the crankshaft is rotating at 1000 rev/min and is in the position show in Fig. 9.12, determine:

(a) the velocity of the piston A;
(b) the magnitude and direction of the velocity of point C, which is 0.15 m from A.

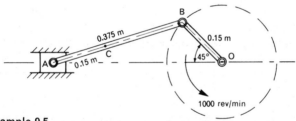

9.12 Example 9.5

Solution

(a) As the crank rotates about O the piston at A is constrained to move along the axis of the cylinder.

The instantaneous velocity of B, v_b, can be obtained from the crank speed given. Applying

$$v = \omega r \qquad \text{(equation (9.3))}$$

gives
$$v_b = \left(\frac{1000}{60} \times 2\pi\right) \times 0.15$$
$$= 15.7 \text{ m/s}$$

This velocity of B relative to O is at right angles to the crank OB. Similarly, as AB is a rigid link, the velocity of A relative to B must be perpendicular to the connecting rod AB. The magnitude of this velocity is unknown.

The velocity vector diagram, Fig. 9.13, can now be drawn.

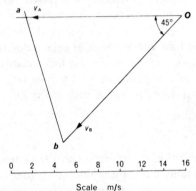

9.13 Velocity vector diagram—Example 9.5

The diagram is constructed as follows:

Starting from O, vector Ob is drawn, to scale, to represent the velocity of B, v_b. From b a line is drawn at right angles to AB to represent the velocity of A relative to B, and a line is drawn through O parallel to the cylinder axis to represent the piston velocity, v_a. These two lines meet at a and vector Oa then represents the velocity of the piston. From the velocity diagram:

$$\text{Velocity of piston} = Oa$$
$$= 14.2 \text{ m/s}$$

(b) As C is on the connecting rod AB its velocity relative to B must be at right angles to the link AB. Thus, the motion of C relative to B is in the same direction as the velocity of A relative to B and so c must lie on ab.

As the link AB can only have one value of angular velocity, then

$$\omega = \frac{ab}{AB} = \frac{cb}{CB}$$

$$\therefore cb = \frac{CB}{AB} \cdot ab$$

$$= \frac{0.225}{0.375} \cdot ab$$

$$= \tfrac{3}{5} \cdot ab$$

Angular motion 171

This enables *c* to be positioned on the velocity diagram, as shown on Fig. 9.14. The absolute velocity of C is then given by vector **Oc**.

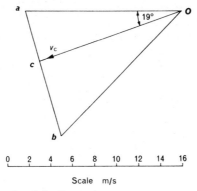

9.14 Absolute velocity of C—Example 9.5

From the vector diagram the velocity of C is 13.7 m/s in a direction making an angle of 19° with the axis of the piston.

Problems

1. A cyclist accelerates from a speed of 12 km/h to a speed of 20 km/h in a time of 20 s while descending an incline. Determine:

(a) the angular acceleration of the road wheels if their diameter is 0.66 m;
(b) the distance travelled in this time.

2. A shaft of diameter 0.30 m is rotating at 2000 rev/min. Determine:

(a) its angular velocity;
(b) the linear speed of a point on the circumference of the shaft;
(c) the angular acceleration required to increase the shaft speed to 3000 rev/min in a period of 50 s;
(d) the linear acceleration of a point on the circumference of the shaft.

3. A 0.375 m diameter wheel is uniformly accelerated from rest such that after 30 revolutions its peripheral speed is 15 m/s. Calculate this angular acceleration and the time taken to attain this speed.

4. A car accelerates from rest to 40 km/h in 8 s. If the wheels are 0.55 m diameter, calculate:

(a) the angular acceleration of the road wheels;
(b) the number of revolutions made;
(c) the distance travelled during the acceleration.

172   Mechanical engineering science

5. A flywheel rotating at 450 rev/min is uniformly retarded at a rate of 1.70 rad/s² for a time of 10 s. Calculate:

(a) the speed of rotation at the end of the retardation;
(b) the number of revolutions made in this time;
(c) the linear retardation of a point on the circumference of the flywheel, whose diameter is 0.64 m.

6. A 0.45 m diameter drum rotating at 300 rev/min is brought to rest in 12 s with a uniform deceleration. Calculate:

(a) the initial linear speed of a point on the drum;
(b) the angular deceleration;
(c) the angle rotated through in this time.

7. A haulage drum of diameter 0.7 m is rotating at 240 rev/min when a brake is applied. The drum is brought to rest in 18 s with a uniform deceleration. Determine:

(a) the angular deceleration of the drum;
(b) the length of rope wound onto the drum during the time of 18 s.

8. The speed of a 2.5 m diameter flywheel is increased from 20 rev/min to 200 rev/min in 45 s. Calculate:

(a) the angular acceleration in rad/s²;
(b) the number of revolutions made by the wheel during the 45 s period.

9. Calculate the centripetal acceleration of a point on the circumference of a 0.40 m diameter wheel which is rotating at 250 rev/min.

10. What is the centripetal acceleration of a point on the periphery of a car tyre of diameter 0.48 m when the speed of the vehicle is 20 km/h? At what vehicle speed will the value of the centripetal acceleration be doubled?

11. A steel wire has a diameter of 0.30 mm and a length of 1 m. A mass of 1.2 kg is attached to the end of the wire and whirled in a circular path in a horizontal plane. If the yield strength of the wire is 300 MN/m² what is the maximum speed at which the mass can be rotated if there is to be a safety factor of 1.6 based upon the yield strength?

12. A mass of 1.2 kg is attached to the end of a cord having a length of 0.88 m and a breaking strength of 440 N. If the mass is rotated in a vertical circle, calculate:

(a) the maximum tension in the cord when the mass is rotating at 80 rev/min;
(b) the speed of rotation at which the cord will break.

Angular motion

13. Masses of 0.8 kg and 1.3 kg are fixed to the ends of arms having lengths of 100 mm and 60 mm respectively. The arms are inclined at 90° to each other and this system is rotated with the arms lying in a plane perpendicular to the axis of rotation. Determine the position and magnitude of a mass which is positioned at a radius of 90 mm such that the system is balanced.

14. Masses of 5 kg, 8 kg, and 10 kg are fixed to a rotating plate at radii of 0.5 m, 0.4 m, and 0.6 m respectively. The angle between the 5 kg and 8 kg masses is 80° and that between the 8 kg and 10 kg masses is 70°, measured in the same direction. Determine:

(a) the resultant force acting on the shaft of the plate when the system is rotating at 200 rev/min;

(b) the position at which a mass of 12 kg should be positioned to balance the system.

15. A disc revolves at 150 rev/min and carries two masses, one of 3 kg at a radius of 0.32 m and the other of 4.2 kg at a radius of 0.4 m. The angle between the radii to the masses is 130°. It is proposed to balance the system by attaching a mass of 4 kg to the disc. Determine the angular position and radius at which this mass should be fixed.

16. The crank OA of an engine mechanism is shown in Fig. 9.15. The crank is 80 mm long and rotates clockwise at 1500 rev/min. The connecting rod AB is 180 mm long. Determine the position shown:

(a) the velocity of the piston B;
(b) the velocity of point C, on the connecting rod, 100 mm from B;
(c) the instantaneous angular velocity of the connecting rod.

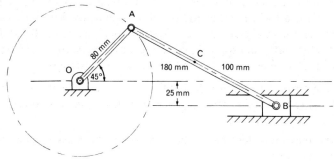

9.15

174 Mechanical engineering science

17. A lorry with wheels of external diameter 0.85 m is travelling along a level road at 70 km/h. If the wheels are not slipping, determine:
(a) their angular velocity;
(b) the velocities of the points on the circumference of the wheel which are at the ends of vertical and horizontal diameters.

18. In a slider crank mechanism the crank AB is 0.25 m long and the connecting rod BC is 1.2 m long. The slider C is in line with the crank centre A. The crank rotates at 500 rev/min. Determine the velocity of the slider when the crank is at 30°, 60°, and 90° respectively from the inner dead centre position.

19. Fig. 9.16 shows a four bar mechanism in which A and D are fixed points. Determine the linear velocity of C and the angular velocities of links BC and CD when the link AB is rotating clockwise at 100 rev/min.

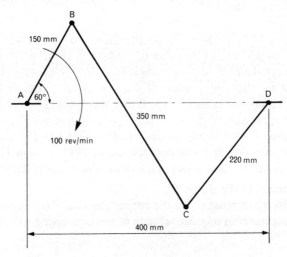

9.16

(Hint: A and D are stationary points and therefore have the same position on the velocity diagram.)

10

Measurement of temperature and pressure

10.1 Measurement of temperature

The measurement of temperature, or degree of hotness, of a substance can be achieved in many ways, although the most common means is by a mercury (or alcohol) in glass thermometer. However, the most important feature in the measurement of temperature is that the property used to determine the temperature is clearly recognisable and that it is capable of repetition. Amongst the properties of a substance which are used to determine temperature are:

(1) the variation in size of a substance, i.e. change in length, area or volume.
(2) the variation in pressure of an enclosed gas.
(3) the change of state of a substance. This property is used to establish points on the International Temperature Scale.
(4) the variation in colour of a substance.
(5) variation in electrical resistance.
(6) variation in radiation from a surface of a body.

10.2 Mercury (or alcohol) in glass thermometers

The mercury-in-glass thermometer is the most widely used method for the determination of temperature. It consists of a capillary tube (fine-bore glass tube) at the lower end of which is fused a glass bulb as shown in Fig. 10.1. The tube and bulb are filled with mercury (or alcohol) and the end sealed when the temperature is above the designed operating temperature of the thermometer.

On cooling, the mercury contracts more than the glass and the level

176 Mechanical engineering science

10.1 Mercury in Glass Thermometer

in the capillary tube therefore falls. The scale is etched on after the glass has fully shrunk.

Mercury is clearly visible and has a fairly uniform rate of expansion. It does not wet the glass and has a range from $-38.86°C$ (freezing point) to $356.7°C$ (boiling point). However, the upper limit of mercury in glass thermometers can be extended to $500°C$ if nitrogen is introduced into the thermometer above the mercury.

Alcohol in glass thermometers is frequently used to measure temperatures lower than $-38°C$ as the freezing point of alcohol is $-113°C$. To enable the temperature to be recorded the alcohol must be dyed (frequently red). With the exception of cost, the use of alcohol has no other advantage over the use of mercury.

Fluid in glass thermometers can produce errors in the temperature recording due to the slow cooling rate of glass and a zero error of $0.1°C$ can be obtained upon cooling from $100°C$.

10.3 The Beckman thermometer

When a small variation in temperature occurs, such as when using a bomb calorimeter, it is vital that this variation can be accurately measured and the Beckman Thermometer is designed for this purpose.

Measurement of temperature and pressure

It can only be used to measure a temperature variation, since it has no fixed points. The range, which is usually 6 K, is calibrated in intervals of 1/100th of a degree.

A diagram of this thermometer is given in Fig. 10.2.

10.2 The Beckman Thermometer

A long capillary tube has a bulb fused at its lower end and a reservoir at its upper end. The capillary tube is surrounded by a hollow glass tube into which the scale is fitted. The mercury level in the capillary tube must be set prior to use. If a temperature increase is to be measured the mercury is set low on the scale while if a temperature decrease is to be measured then the mercury is set high on the scale. To adjust the mercury level the following procedure is adopted.

If the level is too low the bulb is heated so that the level in the capillary tube rises to the level of the reservoir. The thermometer is then cooled so that mercury is pulled from the reservoir into the capillary tube. A light tap on the thermometer at the reservoir level will separate the mercury again and leave a higher level in the capillary tube.

If the level is too high the bulb is again heated so that the mercury in the capillary tube connects up with the mercury in the reservoir. Further heating will drive mercury from the capillary tube into the reservoir and a light tap, as before, will result in a lower level in the capillary tube.

10.4 The gas thermometer

The variation in the volume of a fixed mass of gas at constant pressure, or the variation in the pressure of a fixed mass of gas at constant volume, can be used to measure temperature. This is the principle embodied in the gas thermometer, illustrated in Fig. 10.3.

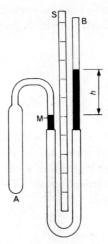

10.3 The Gas Thermometer

A quantity of gas is trapped inside vessel A by the mercury. The glass tube B can be adjusted in height so that the mercury is level with the mark M. The difference in height of the two mercury columns is obtained from the scale S, and by adding the barometric pressure the absolute pressure of the gas is obtained. It will be seen that the volume occupied by the gas is constant.

Before use the thermometer must be calibrated by immersing the bulb firstly in pure melting ice (0°C) and then in the steam of pure boiling water (100°C), both being at standard atmospheric pressure. This will give the pressure of the gas at two known temperatures and as it is assumed that the pressure of the gas varies linearly with temperature it follows that the thermometer can be used to determine the unknown temperature of a liquid or gas.

The gas normally used with this thermometer is hydrogen, which enables the apparatus to be used over a wide temperature range. For very low temperatures helium is used, while for very high temperatures the hydrogen is replaced by nitrogen. When the temperature is such that the glass container would melt it is replaced with one of platinum or platinum–iridium alloy.

It will be apparent that this type of thermometer is cumbersome to handle and it is therefore usually used as a standard for the calibration of other thermometers. For very accurate work, corrections must be made for the variation in the volume of the bulb with changes in the temperature and pressure.

The thermometer can also be used on the constant pressure principle by adjusting the glass tube B so that the two mercury columns are always level.

10.5 Thermocouples

If two wires of dissimilar metals are joined together and their free ends are connected across a galvanometer, as shown in Fig. 10.4, it will be found that a variation in temperature at the junction of the wires produces a deflection of the galvanometer. This arrangement of wires is known as a thermocouple.

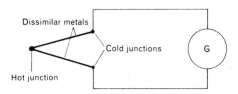

10.4 Thermocouple

The galvanometer deflection indicates the presence of an electric current and as the temperature of the wire junction (hot junction) increases so the deflection of the galvanometer also increases. The current arises:

(*a*) from the variation in temperature at the junction of dissimilar metals (Peltier effect);
(*b*) from the variation in temperature between the hot and cold junctions (Thompson effect).

The thermocouple is used by positioning the hot junction at the position where the temperature is required. For accurate work the cold junction should be maintained at the calibration temperature. The galvanometer can be calibrated to read temperature directly but usually a calibration graph relating the output voltage to the temperature is used.

The most common metals used for the wires are:

(*a*) Chromel–Alumel,
(*b*) Iron–Constantan,
(*c*) Copper–Constantan.

If thermocouples are connected in series for greater sensitivity the arrangement is known as a thermopile.

10.6 The radiation pyrometer

With this type of pyrometer, as opposed to the Optical Pyrometer (Section 10.7), radiant thermal energy is focused upon the hot junction of a thermocouple. A concave mirror, which must be kept clean, is used to focus the radiant energy. A typical arrangement is shown in Fig. 10.5.

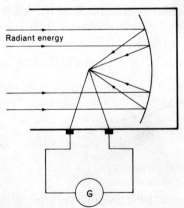

10.5 The Radiation Pyrometer

10.7 The optical pyrometer

When the temperature of certain metals is increased sufficiently they emit light and, as a particular colour occurs at a definite temperature, this can be used to determine a required temperature. The colour variation proceeds from a dull red through to a brilliant white.

The optical pyrometer is based on this property and the high temperature body is viewed through a telescope which incorporates a filter for the protection of the eye.

The most common optical pyrometer is the disappearing filament pyrometer. This operates on the principle of matching the colour of a filament through which an electric current is being passed with the colour of the body whose temperature is required. When a current is passed through a filament the temperature of the filament, and hence its colour, depends upon the value of the current. Thus, with this pyrometer, a filament lamp is positioned in the field of vision of the telescope. The telescope is focused on the body whose temperature is required and the current through the filament then adjusted so that the colour of the filament is identical with the colour of the body, i.e. so that the filament disappears optically. The temperature is obtained from a meter in the filament circuit.

It will be apparent that this type of pyrometer is used for high temperature work and if the entry of light into the pyrometer is restricted, or any light absorbing gases are present between the body and the pyrometer, then errors may occur.

With another form of optical pyrometer the colour of the body is matched with the colour of a filter and errors can arise due to differences in colour interpretation which can occur between persons.

10.8 The resistance thermometer

The electrical resistance of a metal wire is a fundamental property of that wire and the variation of this resistance with temperature is given by

$$R_t = R_0(1+\alpha_0 t)$$

where R_t is the resistance at $t\,°C$
R_0 is the resistance at $0°C$
and α_0 is the temperature coefficient of resistance of the material from, and at, $0°C$.

The resistance thermometer is based upon this variation in electrical resistance. The measuring resistor, which is usually a platinum winding, is connected into a Wheatstone Bridge circuit, as shown in Fig. 10.6.

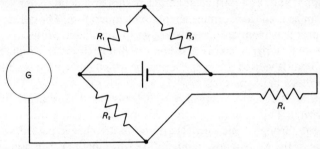

10.6 The Optical Pyrometer

The resistances R_1 and R_2 are fixed resistances, that are usually equal, while R_3 is a variable resistor and R_4 the measuring resistor. In use, the value of R_3 is adjusted to produce zero deflection of the galvanometer. Hence R_4 can be determined.

The thermometer is calibrated by determing the value of the measuring resistor at 0°C (in melting ice) and 100°C (in boiling water at standard atmospheric pressure). This enables R_0 and α_0 to be determined and if the resistance R_t is then obtained at an unknown temperature, this temperature can be calculated.

A high degree of accuracy can be obtained with this thermometer. However, the measuring resistor is fragile and although it is usually enclosed in a protecting tube it cannot be used if excessive vibration or shock conditions are present.

10.9 Measurement of pressure

The pressure of a liquid or gas is a measure of the force exerted per unit area. The most frequently used methods of measuring pressure are:

(*a*) the height of a liquid column as in a Fortin barometer or a water manometer;
(*b*) the variation in volume of a closed cell, as in the Aneroid barometer;
(*c*) the pressure gauge;
(*d*) the action of pressure against a spring, as in the engine indicator diagram.

Measurement of temperature and pressure

10.10 The liquid column barometer

The height of a column of liquid, usually mercury, which can be supported by the prevailing atmospheric pressure is used to evaluate that atmospheric pressure.

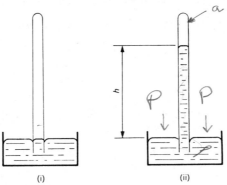

10.7 The Liquid Column Barometer

Suppose a glass tube which is sealed at one end is placed, with its sealed end uppermost, in a vessel containing a liquid, as shown in Fig. 10.7(i). As the air pressure in the tube will be equal to the atmospheric pressure, the liquid level in the tube will be the same as that in the vessel. However, if the tube is now evacuated the liquid level will rise to a height h, as shown in Fig. 10.7(ii). The height h will be such that the pressure on the liquid in the vessel due to the mass of liquid in the tube is equal to the atmospheric pressure, P, acting on the liquid in the vessel.

If the area of the tube is a and the density of the liquid ρ, then:

$$\text{Mass of liquid in tube} = \text{density} \times \text{volume}$$
$$= \rho a h$$
$$\text{Gravitational force on liquid in tube} = 9.81 \times \rho a h$$
$$\text{Pressure on liquid in vessel} = \frac{9.81 \times \rho a h}{a}$$
$$P = 9.81 \rho h$$

As the density of a liquid is reasonably constant it follows that the height of the column, h, is proportional to the atmospheric pressure, P. If the atmospheric pressure falls then the height of the liquid column will also fall. Conversely, if the atmospheric pressure increases then the height of the liquid column will also rise.

Obviously the height to which the liquid column rises depends upon the liquid used. Let us suppose that this is water.

$$\text{Density of water} = 1 \text{ Mg/m}^3 = 10^3 \text{ kg/m}^3$$

Then for standard atmospheric pressure of 101.3 kN/m²

$$101.3 \times 10^3 \text{ [N/m}^2\text{]} = 9.81 \text{ [m/s}^2\text{]} \times 10^3 \text{ [kg/m}^3\text{]} \times h$$

$$\therefore h = \frac{101.3}{9.81} \frac{\text{[N/m}^2\text{]}}{\text{[N/m}^3\text{]}}$$

$$= 10.33 \text{ m}$$

Obviously a column of water 10.33 m high is impracticable to use and so, to reduce this, mercury is used.

$$\text{Density of mercury} = 13.6 \text{ Mg/m}^3$$

Then,

$$101.3 \times 10^3 \text{ [N/m}^2\text{]} = 9.81 \text{ [m/s}^2\text{]} \times 13.6 \times 10^3 \text{ [kg/m}^3\text{]} \times h$$

$$h = \frac{101.3}{9.81 \times 13.6}$$

$$= 0.76 \text{ m} = 760 \text{ mm}$$

Thus, we say that the atmospheric pressure is 760 mm of mercury.

The Fortin Barometer uses this principle to enable the atmospheric pressure to be determined. In addition to the mercury column, a scale and reservoir level adjusting screw are fitted. A thermometer is usually fitted to the front of this barometer.

10.11 The aneroid barometer

The basic component of an aneroid barometer, shown in Fig. 10.8, is a partially evacuated corrugated cell. The cell is made from thin metal

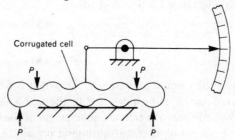

10.8 Principle of the Aneroid Barometer

and is circular and hollow. The external surface of the cell is subjected to atmospheric pressure and if this varies then the surface of the cell will move. This movement is magnified and recorded by a pointer moving over a scale. The scale is usually circular and calibrated in mm of mercury, only the relevant part of the scale on either side of normal atmospheric pressure being shown.

10.12 The manometer

A manometer is used to obtain the difference between the pressure of a gas or vapour and atmospheric pressure. It consists of a U-tube containing liquid, and a measuring scale. The liquid used depends upon the pressure difference to be measured. If the difference is fairly large then mercury is used but for small differences in pressure water is used.

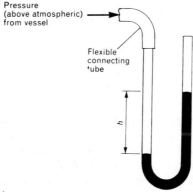

10.9 The Manometer

Suppose a vessel containing a gas at a pressure above atmospheric pressure is connected to the manometer as shown in Fig. 10.9. The additional pressure of the gas will cause the mercury levels to move so that their difference in heights is h. Then,

Pressure of gas = atmospheric pressure
+ pressure due to liquid column of height h

If the pressure of the gas in the vessel were less than atmospheric pressure then the mercury level in the left-hand column would rise above that in the right-hand column. Thus, in such a case, h would be negative so giving,

Pressure of gas = atmospheric pressure
− pressure due to liquid column of height h

186 Mechanical engineering science

The difference between the actual pressure (absolute pressure) of a gas and the atmospheric pressure is known as the gauge pressure. If the gas pressure is less than atmospheric pressure the gauge pressure is negative. Thus,

Absolute pressure = atmospheric pressure + gauge pressure

10.13 The inclined manometer

When the pressure difference to be measured is very small an inclined manometer, Fig. 10.10, is used to achieve a greater accuracy.

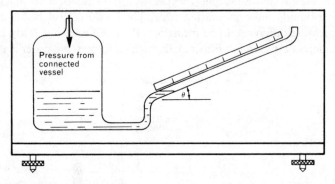

10.10 The Inclined Manometer

The liquid used is water and the manometer comprises a large reservoir which is connected to a manometer tube having an inclination θ to the horizontal. The apparatus is equipped with levelling screws to ensure that the angle θ is correctly set. The scale fixed to the manometer tube is calibrated in terms of the vertical height and because of the extra length this scale can be more finely calibrated.

The reservoir is connected to the vessel containing the gas whose pressure is required. This will cause a movement in the liquid levels of the manometer, but by having a large reservoir the movement of the liquid level in this becomes so small, compared with the movement in the tube, that it can be neglected. Thus, the change in pressure is recorded directly on the manometer tube.

10.14 The Bourdon pressure gauge

The manometer is not suitable for measuring large gauge pressures because of the length of column required. In such instances a Bourdon pressure gauge is used, a diagram of which is given in Fig. 10.11.

10.11 The Bourdon Pressure Gauge

This consists of a hollow tube having an elliptical cross-section which is bent into an arc. One end of this tube is sealed and the other end connected to the pressure to be measured. With increasing pressure the tube tends to straighten out and this movement is transmitted via a rack and pinion to a pointer moving over a calibrated scale.

This type of gauge can be used to measure pressures below atmospheric pressure, when it is known as a vacuum gauge, as well as for very high pressures.

10.15 The engine indicator

The methods so far described for the measurement of pressure can only be usefully employed when the pressure is reasonably constant. During the operating cycle of an engine the pressure within the cylinder varies so rapidly over a considerable range that it is necessary to use special apparatus to record the pressure variations. This apparatus is known as an engine indicator and a diagram is given in Fig. 10.12.

The indicator consists of a cylinder fitted with a coupling nut to enable the equipment to be connected to a gas tap on the cylinder of the test engine. When the regulator valve is opened the pressure acts on a

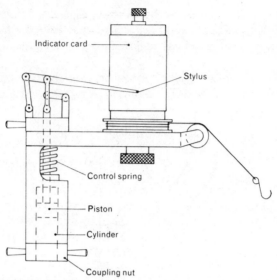

10.12 An engine indicator

piston which is free to slide within the cylinder. The movement of this piston is transmitted, via a control spring of known stiffness, to a linkage, as shown in Fig. 10.12. This linkage is mounted on a freely rotatable collar at the upper end of the cylinder and the end of the linkage is fitted with a stylus. As the piston moves so also does the stylus, its movement being magnified by the linkage. The stylus can be brought into contact with a drum which is mounted on a plate, extending from the top of the cylinder. The drum is free to oscillate on a central spindle, its return being controlled by a spring. An indicator card is attached to the drum by means of a spring clip. The rotation of the drum is brought about via a cord wrapped around a channel at the lower end of the drum. A hook is attached to the end of the cord and when this hook is connected to a suitable reciprocating component on the engine the piston movement within the engine is reproduced.

The operating procedure is as follows.

Attach an indicator card to the drum. This card has a special surface so that a black line is produced as the stylus passes over it.

Connect the hook to the reciprocating mechanism so that the drum oscillates and so obtain the horizontal component on the indicator diagram.

Open the gas tap on the engine. This will cause the stylus to move vertically up and down. Rotate the collar until the stylus touches the

Measurement of temperature and pressure

indicator card so producing the indicator diagram. This only takes a very short time. The stylus is then rotated away from the drum, the gas tap is turned off and the hook removed from the reciprocating mechanism.

The height of the indicator diagram depends upon the stiffness of the control spring. With higher engine loads a spring of greater stiffness is used.

10.16 Pressure–volume diagram. Work done

Suppose a volume of gas V_1 at a pressure P is contained in a cylinder by a piston of area A. The force on the piston will be $P \times A$. If the piston is moved by the gas through a distance L such that the volume becomes V_2, the pressure remaining constant throughout then the force on the piston will also remain constant.

$$\text{Word done by gas} = \text{force on piston} \times \text{distance moved}$$
$$= PA \times L$$

But AL is equal to the increase in volume, i.e. $V_2 - V_1$

$$\therefore \text{Work done by gas} = P(V_2 - V_1)$$

Let us now look at the pressure volume diagram for this expansion, shown in Fig. 10.13. It will be seen that the shaded area under the graph is also equal to $P(V_2 - V_1)$.

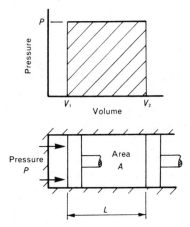

10.13 Pressure–volume diagram for constant pressure expansion

Thus, for a constant pressure expansion the area under the pressure–volume diagram is equal to the work done by the gas.

If the pressure varies during the expansion of the gas the work done is still equal to the area under the pressure–volume diagram. Consider the expansion shown in Fig. 10.14 from an original pressure P_1 and

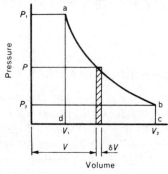

10.14 Pressure–volume diagram—Work done

volume V_1 to conditions P_2 and V_2 respectively. At some stage during this expansion process let the instantaneous pressure and volume be P and V respectively. Suppose the volume now increases by a small increment δV.

Work done during incremental expansion $= P \cdot \delta V$

$\qquad\qquad\qquad\qquad\qquad\qquad\quad\;\; =$ area of strip (shaded)

The total work done during the expansion from P_1 to P_2 is therefore equal to the sum of all such incremental strips, i.e.

$$\text{Total work done} = \sum_{V_1}^{V_2} P \cdot \delta V$$

$$= \text{area abcd}$$

We have so far dealt with an expansion process and the work done by the gas. A compression process is the reverse of expansion and the work done is therefore negative, i.e. work is done on the gas to compress it.

10.17 The indicator diagram

A typical indicator diagram for a 4-stroke engine is shown in Fig. 10.15. The expansion starts at a and as a result of ignition the pressure rises to b and then decreases to c. The work done by the expanding gases is therefore represented by area abcef. Following the exhaust and inlet

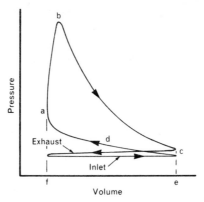

10.15 Indicator diagram—4 stroke cycle

strokes of the engine, compression of the new petrol–air mixture occurs along the line cda. The work done to bring about this compression is represented by area adcef.

Thus,

Total work done during 1 cycle of the engine
$$= \text{area abcef} - \text{area adcef}$$
$$= \text{area abcda}$$

Now the indicated mean effective pressure of an engine is that constant pressure which, if acting for the complete length of stroke would produce the same quantity of work done as is actually achieved during the engine cycle. Thus,

Work done = indicated mean effective pressure
× piston area × stroke
= i.m.e.p. × swept volume

The i.m.e.p. is obtained from the indicator diagram by determining the area abcda. This is usually determined by using a planimeter but if this equipment is not available the mid-ordinate method can be used.

When discussing the engine indicator it was mentioned that the pressure of the gas acts against the action of a control spring of known strength. Thus, if the control spring is marked 25 kN/m² per mm this means that for each mm of vertical movement of the stylus the pressure in the engine cylinder has increased by 25 kN/m². Then,

i.m.e.p. = mean height of diagram × spring calibration

$$= \frac{\text{diagram area}}{\text{diagram length}} \times \text{spring calibration}$$

11

Gases

11.1 Nature of a gas

In a solid the constituent atoms or molecules have a very limited degree of freedom, while in a liquid the molecules are free to move within the boundary confines of the liquid and so are always in continual contact. However, in the case of a gas the molecules move freely and rapidly within the space occupied by the gas. Thus, a gas always fills its containing vessel and collisions continually occur between individual molecules and between the molecules and the walls of the containing vessel. The pressure exerted by a gas is the result of these collisions between the molecules and the vessel walls.

11.2 Kinetic theory of gases

In the preceding section it has been stated that the molecules of a gas are in a state of rapid random movement within the space occupied by the gas. This means that there will be continual impacts between the molecules and the walls of the containing vessel and if a molecule is regarded as perfectly elastic it will rebound with a velocity of the same magnitude, only the direction of motion being changed. Such an impact produces a force on the wall of the vessel and the pressure exerted by a gas is the net sum of these individual forces per unit area, bearing in mind that a molecule is very small. Furthermore, as the molecules rebound from the vessel wall with a velocity of equal magnitude it follows that the kinetic energy (energy of motion) possessed by the enclosed gas remains constant unless the physical conditions of the gas are changed.

Although the molecules are moving freely in all directions there is no physical impact between them. If two molecules are travelling on a collision course then just as impact is imminent repulsion forces are created

Gases

causing deflection of the molecules concerned. This can be compared with the forces of repulsion set up by like magnetic poles.

The kinetic theory of gases also makes the following assumptions:

1. The size of a molecule is very small compared with the distance travelled between encounters with other molecules or impacts with the boundary wall.
2. The volume occupied by the molecules is very small compared with the volume of the containing vessel.
3. The molecules are sufficiently far apart for any mutual forces between them to have no effect on the overall behaviour of the gas.
4. The molecules are perfectly elastic.

11.3 Perfect gas or ideal gas concept

A perfect gas can be defined as one which obeys the gas laws (Sections 11.4–11.7) exactly. In practice no such gas exists, although this ideal state of affairs is approached when the temperature of a gas is considerably above its liquifying temperature. For the temperatures and pressures of the gases which we are concerned with at this stage the concept of an ideal gas enables the behaviour of gases to be appreciated.

11.4 Boyle's law

This law, which is named after Sir Robert Boyle (1627–1691), an Irish scientist, states that:

The volume of a given mass of gas at constant temperature is inversely proportional to its absolute pressure.

$$\text{Volume } (V) \propto \frac{1}{\text{pressure } (P)}$$

i.e. $PV = \text{constant}$

If suffices 1 and 2 denote the initial and final conditions, respectively, then

$$P_1 V_1 = P_2 V_2 \qquad (11.1)$$

Changes in a gas which occur at constant temperature are known as isothermal changes.

Example 11.1

A gas occupies a volume of 0.6 m³ at a pressure of 135 kN/m². If the gas is expanded at constant temperature so that its volume increases to 0.84 m³ what will be the resulting pressure?

Mechanical engineering science

Solution

Applying Boyle's law:

$$P_1 V_1 = P_2 V_2$$

where $P_1 = 135$ kN/m²; $V_1 = 0.6$ m³; $V_2 = 0.84$ m³. Then

$$P_2 = \frac{P_1 V_1}{V_2} = 135 \times \frac{0.6}{0.84}$$

$$= 96.4 \text{ kN/m}^2$$

The resultant pressure of the gas will be 96.4 kN/m².

11.5 Charles' law

This law, named after the French physicist Jacques Charles (1746–1823), is concerned with the variation of volume and temperature of a gas when its pressure remains constant, and can be stated in the form:

The volume of a given mass of gas, at a constant pressure, is directly proportional to its thermodynamic temperature.

$$\text{Volume } V \propto \text{ thermodynamic temperature } T$$

or

$$\frac{V}{T} = \text{constant}$$

The thermodynamic temperature is that measured on the Kelvin (K) scale.

The relationship between thermodynamic temperature and Celsius temperature is:

Thermodynamic temperature (K) = Celsius temperature (°C) + 273.15

For practical purposes, however, it is customary to use

$$K = °C + 273$$

Note that the temperature interval on the Kelvin scale is equal to that on the Celsius scale.

Returning to our relationship above—if suffices 1 and 2 again refer to the initial and final conditions, respectively, then:

$$\frac{V_1}{T_1} = \frac{V_2}{T_2} \qquad (11.2)$$

Example 11.2

A gas is contained in a cylinder at a pressure of 320 kN/m². It occupies a volume of 0.015 m³ at a temperature of 18°C. To what temperature must the gas be cooled, if the pressure remains constant, so that the volume is reduced to 0.0138 m³?

Solution

As the gas is cooled at constant pressure then Charles' law can be applied, i.e.

$$\frac{V_1}{T_1} = \frac{V_2}{T_2}$$

where $\quad V_1 = 0.015 \text{ m}^3; \quad V_2 = 0.0138 \text{ m}^3$

$T_1 = 18°C = (18+273) = 291 \text{ K}$

Then $\quad T_2 = \frac{V_2}{V_1} \times T_1$

$= \frac{0.0138}{0.015} \times 291 \text{ K}$

$= 267.7 \text{ K}$

$= 267.7 - 273 = -5.3°C$

The gas must be cooled to $-5.3°C$.

11.6 Dalton's law of partial pressures

This law states that the total pressure of a mixture of gases is equal to the sum of the pressures of each gas, considered separately, assuming it to occupy the same volume as the mixture and to be at the same temperature.

The pressure of each constituent gas is known as the partial pressure of that gas. This partial pressure is independent of the presence of other constituents. Thus,

Total pressure of a gas mixture = sum of the partial pressures of the individual gases

11.7 The characteristic gas equation

There are three properties of state of a gas, namely pressure, temperature, and volume. In Sections 11.4 and 11.5 the laws were stated relating any two quantities, while the third remained constant. We must now consider the case when all three quantities vary.

Consider a mass of gas whose initial pressure, volume, and temperature are P_1, V_1, and T_1 respectively to be subjected to a change of state so that its corresponding final conditions are P_2, V_2, and T_2. This final state could be achieved in a variety of ways, the actual path being quite arbitrary. Let us therefore consider a possible route, comprising two stages.

Let the pressure be changed from P_1 to P_2 while the temperature remains constant at T_1. The volume will change from V_1 to an intermediate value V.

Then as this is a Boyle's law expansion:

$$P_1 V_1 = P_2 V$$

Now consider the pressure to remain constant at P_2 while the temperature is changed from T_1 to T_2 and the volume changes from V to V_2.

This expansion, at constant pressure, is a Charles' law expansion. Thus:

$$\frac{V}{T_1} = \frac{V_2}{T_2}$$

Substituting for V from the previous equation gives:

$$\frac{P_1 V_1}{P_2 T_1} = \frac{V_2}{T_2}$$

$$\frac{P_1 V_1}{T_1} = \frac{P_2 V_2}{T_2}$$

or, generally
$$\frac{PV}{T} = \text{constant} \qquad (11.3)$$

This equation is applicable to a fixed mass of gas. Let this mass be m. Equation (11.3) can then be written

$$\frac{PV}{T} = m \times \text{constant}$$

$$\frac{PV}{T} = mR \qquad (11.4)$$

where R is known as the characteristic gas constant.

Equation (11.4) is known as the characteristic equation of a perfect gas.

If P has units of N/m², V has units of m³, T has units of K, and m has units of kg, then the units of R are

$$\frac{\text{N m}^3}{\text{m}^2 \text{ K kg}} \quad \text{or} \quad \frac{\text{J}}{\text{kg K}}$$

Example 11.3

A volume of 0.025 m³ of oxygen at a pressure of 300 kN/m² and a temperature of 15°C is expanded to a volume of 0.045 m³. If the temperature falls to 3°C calculate the final pressure of the gas.

Solution

Applying
$$\frac{P_1 V_1}{T_1} = \frac{P_2 V_2}{T_2}$$

where $P_1 = 300 \times 10^3$ N/m²; $P_2 =$ final pressure
$V_1 = 0.025$ m³; $V_2 = 0.045$ m³
$T_1 = 15 + 273 = 288$ K; $T_2 = 3 + 273 = 276$ K

Then
$$P_2 = \frac{P_1 V_1}{T_1} \times \frac{T_2}{V_2}$$

$$= 300 \times 10^3 \times \frac{0.025}{288} \times \frac{276}{0.045}$$

$$= 159 \times 10^3 \text{ N/m}^2$$

$$= 159 \text{ kN/m}^2$$

The final pressure of the gas is 159 kN/m².

Example 11.4

Air is drawn into an engine cylinder at a pressure of 96 kN/m² and temperature of 16°C. At the end of the compression stroke the pressure is 1300 kN/m² and the temperature is 150°C. Determine the compression ratio of the engine.

Solution

$$\text{Compression ratio} = \frac{\text{initial volume } V_1}{\text{final volume } V_2}$$

Applying
$$\frac{P_1 V_1}{T_1} = \frac{P_2 V_2}{T_2}$$

for the compression and rewriting:

$$\frac{V_1}{V_2} = \frac{P_2}{P_1} \times \frac{T_1}{T_2}$$

$$= \frac{1300}{96} \times \left(\frac{273+16}{273+150}\right)$$

$$= 9.25$$

The compression ratio is 9.25.

Example 11.5

A cylinder has a volume of 0.11 m³ and contains nitrogen at a pressure of 1720 kN/m² and temperature of 18°C. After some of the gas has been consumed it is found that the pressure has fallen to 1200 kN/m² and the temperature is then 10°C. Determine the volume of gas at s.t.p. that has been drawn off from the cylinder.

Solution

Let the volume of gas consumed at 1720 kN/m² and temperature of 18°C be V m³.

Then,   Volume remaining in cylinder = $(0.11 - V)$ m³

This volume expands to fill the cylinder, i.e. to a volume of 0.11 m³, and as a result its pressure falls to 1200 kN/m² and its temperature drops to 10°C.

Applying $$\frac{P_1 V_1}{T_1} = \frac{P_2 V_2}{T_2}$$

where $P_1 = 1720 \times 10^3$ N/m²; $P_2 = 1200 \times 10^3$ N/m²
$V_1 = (0.11 - V)$ m³; $V_2 = 0.11$ m³
$T_1 = 18 + 273 = 291$ K; $T_2 = 10 + 273 = 283$ K

Then $$\frac{1720 \times 10^3 \times (0.11 - V)}{291} = \frac{1200 \times 10^3 \times 0.11}{283}$$

$$0.11 - V = \frac{1200 \times 0.11 \times 291}{1720 \times 283}$$

$$= 0.079$$

$$\therefore V = 0.031 \text{ m}^3$$

This volume must now be expanded to s.t.p. conditions, i.e. 101.3 kN/m² and 0°C.

Applying
$$\frac{P_1 V}{T_1} = \frac{P_s V_s}{T_s}$$

where $P_s = 101.3 \times 10^3$ N/m²; $T_s = 273$ K;

$V_s =$ volume at s.t.p.

gives
$$\frac{1720 \times 10^3 \times 0.031}{291} = \frac{101.3 \times 10^3 \times V_s}{273}$$

$$V_s = \frac{1720 \times 0.031 \times 273}{291 \times 101.3}$$

$$= 0.494 \text{ m}^3$$

The volume of gas consumed at s.t.p. is 0.494 m³.

Example 11.6

A spherical vessel having a diameter of 2 m contains helium at a pressure of 300 kN/m² and temperature of 115 K. Determine the mass of helium contained in the vessel given that the gas constant R for helium is 2078.6 J/kg K.

Solution

Applying
$$PV = mRT$$

$$m = \frac{PV}{RT}$$

where $P = 300 \times 10^3$ N/m²; $V = \frac{4}{3} \times \pi \times 1^3 = \frac{4\pi}{3}$ m³

$R = 2078.6$ J/kg K; $T = 115$ K

Then,
$$m = \frac{300 \times 10^3 \text{ [N/m}^2\text{]} \times \frac{4\pi}{3} \text{ [m}^3\text{]}}{2078.6 \text{ [J/kg K]} \times 115 \text{ [K]}}$$

$$= 5.25 \text{ kg}$$

The mass of helium in the vessel is 5.25 kg.

11.8 The mole (or mol)

The mole of a gas is defined as the mass of gas equal to the relative molecular mass. Thus, since the relative molecular mass of hydrogen is 2, 1 kg mole of hydrogen is equal to 2 kg of hydrogen. Similarly, 1 kg mole of oxygen is equal to 32 kg of oxygen.

Thus, for a mass m of gas

$$m = nM \qquad (11.5)$$

where $\quad n$ = number of moles

and $\quad M$ = molecular mass of gas

11.9 Avogadro's hypothesis

Avogadro's hypothesis states that:

Equal volumes of different gases, at the same temperature and pressure, contain equal numbers of molecules.

Thus, at constant pressure and temperature, V/n is constant for all gases.

11.10 Universal gas constant

By the characteristic equation for a perfect gas (equation (11.4))

$$\frac{PV}{T} = mR$$

Substituting for m from equation (11.5)

$$\frac{PV}{T} = nMR \quad \text{or} \quad \frac{PV}{nT} = MR$$

Now, by Avogadro's hypothesis, V/n is constant for all gases if the pressure and temperature are constant.

Thus, PV/nT is constant for all gases.

This constant is called the universal gas constant, R_0. Hence

$$MR = R_0$$

$$R = \frac{R_0}{M}$$

Equation (11.4) can now be written

$$\frac{PV}{T} = m\frac{R_0}{M} \qquad (11.6)$$

The value of R_0 is 8314.3 J/kg mole K.

11.11 Volume of one mole of a gas

Since
$$\frac{PV}{T} = m\frac{R_0}{M}$$

and for 1 mole of gas $m = M$, then

$$\frac{PV}{T} = R_0$$

$$V = \frac{R_0 T}{P}$$

Thus, at s.t.p. conditions (273 K and 101.3 kN/m²) the volume V of 1 mole of any gas is

$$V = \frac{8314.3 \times 273}{101.3 \times 10^3}$$

$$= 22.4 \text{ m}^3$$

Example 11.7

0.75 kg of a gas having a relative molecular mass of 28 and pressure of 392 kN/m² occupy a cylinder at a temperature of 12°C. Determine:

(a) the gas constant;
(b) the volume of the cylinder.

(Universal gas constant = 8314.3 J/kg mol K.)

Solution

(a) From Section 11.10

$$R = \frac{R_0}{M}$$

Thus,
$$R = \frac{8314.3}{28}$$

$$= 296.9 \text{ J/kg K}$$

(b) Since $\quad PV = mRT \quad$ (equation (11.4))

$$V = \frac{mRT}{P}$$

where

$m = 0.75$ kg; $\quad R = 296.9$ J/kg K
$T = 12 + 273 = 285$ K; $\quad P = 392 \times 10^3$ N/m²

Thus, $$V = \frac{0.75 \times 296.9 \times 285}{392 \times 10^3}$$

$$= 0.162 \text{ m}^3$$

The gas constant is 296.9 J/kg K and the volume of the cylinder is 0.162 m³.

Example 11.8

A vessel of total volume 0.7 m³ contains a mixture of oxygen and nitrogen. The total pressure of the mixture is 530 kN/m² at a temperature of 24°C. If the mass of oxygen present is 0.8 kg calculate:

(a) the partial pressure of each gas;
(b) the mass of nitrogen present.

(Relative molecular mass of oxygen = 32;
Relative molecular mass of nitrogen = 28;
Universal gas constant $R_0 = 8314.3$ J/kg mol K.)

Solution

(a) We know that 0.8 kg of oxygen is occupying a volume of 0.7 m³ at a temperature of 24°C. The pressure exerted by the oxygen can therefore be calculated from

$$\frac{PV}{T} = m \frac{R_0}{M}$$

$$P = \frac{mR_0T}{MV}$$

$$= \frac{0.8 \text{ [kg]} \times 8314.3 \text{ [J/kg mol K]} \times (273+24) \text{ [K]}}{32 \text{ [/mol]} \times 0.7 \text{ [m}^3\text{]}}$$

$$= 88\,190 \text{ N/m}^2$$

$$= 88.19 \text{ kN/m}^2$$

By Dalton's law of partial pressures (Section 11.6)

Total pressure of mixture = oxygen pressure + nitrogen pressure

∴ Nitrogen pressure = 530 − 88.19

$$= 441.81 \text{ kN/m}^2$$

(b) We now know that the nitrogen at a pressure of 441.81 kN/m² and temperature of 24°C occupies a volume of 0.7 m³. Hence the mass of nitrogen in the vessel can be obtained by applying

$$PV = \frac{mR_0T}{M} \quad \text{or} \quad m = \frac{PVM}{R_0T}$$

where

$$P = 441.81 \times 10^3 \text{ N/m}^2; \quad V = 0.7 \text{ m}^3; \quad M = 28$$
$$R_0 = 8314.3 \text{ J/kg mol K}; \quad T = 273 + 24 = 297 \text{ K}$$

$$m = \frac{441.81 \times 10^3 \times 0.7 \times 28}{8314.3 \times 297}$$

$$= 3.5 \text{ kg}$$

The mass of nitrogen present in the vessel is 3.5 kg.

11.12 Specific heat capacities

The specific heat capacity of a substance is the quantity of thermal energy required to give a unit mass of the substance a unit temperature rise.

In the case of a gas the increase in temperature could take place at constant pressure, at constant volume or with both the pressure and volume varying. Hence, there could be an infinite number of specific heat capacities depending upon the conditions under which the temperature rise occured. It is therefore important when quoting the specific heat capacity of a gas to state the pressure and volume conditions applicable. Two specific heat capacities are defined:

(a) Specific heat capacity at constant volume, c_v. This is the quantity of thermal energy which must be transferred in order to increase the temperature of unit mass of the gas by one degree with the volume remaining constant.

(b) Specific heat capacity at constant pressure, c_p. This is the quantity of thermal energy which must be transferred in order to increase the temperature of unit mass of the gas by one degree with the pressure remaining constant.

The units for c_v and c_p are therefore J/kg K.

These specific heat capacities vary with temperature and usually an average value is used for the temperature range involved.

Typical values are given in Table 3 (see p. 204).

Table 3. Average values for the specific heat capacities of gases

GAS	SPECIFIC HEAT CAPACITY kJ/kg K	
	c_p	c_v
Air	1.005	0.712
Carbon monoxide	1.046	0.759
Carbon dioxide	0.837	0.653
Hydrogen	14.486	10.383
Oxygen	0.921	0.67
Nitrogen	1.046	0.754

11.13 Relationship between specific heat capacities and gas constant

By considering the heat energy transfer required to cause a temperature increase of a gas at (*a*) constant volume and (*b*) constant pressure we can obtain a relationship between the specific heat capacities of a gas and the gas constant.

(*a*) HEATING AT CONSTANT VOLUME

Consider a mass m of gas to be contained within a cylinder by a piston such that the pressure is P_1, the volume V_1, and the temperature T_1. Suppose a transfer of heat energy now occurs so that the pressure is increased to P_2 and the temperature to T_2 while the volume remains constant at V_1, as shown in Fig. 11.1.

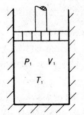
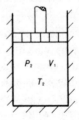

11.1 Constant volume heating of a gas

Then,

Energy transferred = mass × specific heat capacity at constant volume × temperature rise

$$= mc_v(T_2 - T_1) \qquad (11.7)$$

Now, as this energy transfer takes place at constant volume no external work is done. Thus, all the energy transferred is absorbed by the gas as internal energy.

$$\therefore \text{ Increase of internal energy of gas} = mc_v(T_2 - T_1) \quad (11.8)$$

(*b*) HEATING AT CONSTANT PRESSURE

Let us again start with mass m of gas at conditions of state P_1, V_1, and T_1, respectively. In this instance assume an energy transfer occurs such that the pressure remains constant at P_1, while the volume increases to V_2 and the temperature to T_2.

This expansion is depicted in Fig. 11.2.

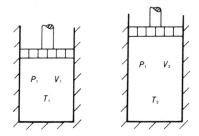

11.2 Constant pressure heating of a gas

Energy transferred

$\quad\quad\quad$ = mass × specific heat capacity at constant pressure × temperature rise

$$= mc_p(T_2 - T_1) \quad (11.9)$$

In addition to increasing the temperature of the gas from T_1 to T_2 some of the energy transferred is absorbed by the external work done by the gas. Hence, the energy transfer required at constant pressure will obviously be greater than that required at constant volume.

$$\text{External work done by gas} = P_1(V_2 - V_1)$$

But from the characteristic equation of a gas (equation (11.4)) $PV = mRT$. Thus,

$$P_1 V_1 = mRT_1 \quad \text{and} \quad P_1 V_2 = mRT_2$$

$$\therefore \text{ External work done by gas} = mR(T_2 - T_1)$$

206 Mechanical engineering science

Then,

$$\text{Increase of internal energy of gas} = \text{energy transferred} - \text{external work done}$$

$$= mc_p(T_2-T_1) - mR(T_2-T_1) \quad (11.10)$$

Now Joule's law, which is named after James Prescott Joule (1818–1889), an English physicist, states that:

The internal energy of a gas is a function of temperature only and is independent of any changes occurring in the pressure and volume.

Further discussion on the internal energy of a gas will not be pursued at this stage. Such discussions will, however, be an important part of further studies in Thermodynamics.

Thus, as expressions (11.8) and (11.10) are functions of temperature only they must, by Joule's law, be equal, i.e.

$$mc_v(T_2-T_1) = mc_p(T_2-T_1) - mR(T_2-T_1)$$

Hence,
$$c_v = c_p - R$$

or
$$c_p - c_v = R \quad (11.11)$$

Example 11.9

A vessel having a volume of 8 m³ contains 0.338 kg of hydrogen at a pressure of 50 kN/m² and temperature of 22°C.

Given that the specific heat capacity at constant volume c_v is 10 390 J/kg K determine the specific heat capacity of hydrogen at constant pressure.

Solution

Applying the characteristic gas equation

$$PV = mRT$$

we can determine the gas constant R for hydrogen, since

$$P = 50 \times 10^3 \text{ N/m}^2; \quad V = 8 \text{ m}^3$$

$$m = 0.338 \text{ kg}; \quad T = 273 + 22 = 295 \text{ K}$$

Then $\quad R = \dfrac{PV}{mT}$

$$= \dfrac{50 \times 10^3 \,[\text{N/m}^2] \times 8 \,[\text{m}^3]}{0.338 \,[\text{kg}] \times 295 \,[\text{K}]}$$

$$= 4012 \text{ J/kg K}$$

But $\quad R = (c_p - c_v) \quad\quad$ (equation (11.11))

$\therefore\; c_p = R + c_v$

$\quad\quad = 10\,390 + 4012$

$\quad\quad = 14\,400$ J/kg K

The specific heat capacity of hydrogen at constant pressure, c_p, is 14 400 J/kg K.

11.14 Conversion of a volumetric analysis of a gas mixture to a mass analysis

When dealing with problems involving gas mixtures a conversion of a gas analysis by volume to one by mass, or vice versa, is frequently required.

From Avogadro's hypothesis (Section 11.9) we saw that, at constant pressure and temperature, the volume of a gas (V) divided by the number of moles (n) was a constant for all gases. Thus, proportions by molecules are also proportions by volume. Then, from equation (11.5), viz. $m = nM$ it follows that, for a gas,

Proportion by mass = proportion by volume
$\quad\quad\quad\quad\quad\quad\quad\quad\quad\quad \times$ relative molecular mass

Using this relationship a conversion from a volumetric analysis to a mass analysis can be carried out, as illustrated in Example 11.10 below.

Example 11.10

A volumetric analysis of a gas mixture gave the following results

Carbon dioxide (CO_2) 35%;   Methane (CH_4) 20%;
Hydrogen (H_2) 15%;   Nitrogen (N_2) 30%.

Determine the analysis of the gas mixture by mass.

208 Mechanical engineering science

Solution

A tabular solution is preferable for this type of problem.

CONSTITUENT	RELATIVE MOLECULAR MASS (1)	FRACTION BY VOLUME (2)	(1) × (2)	FRACTION BY MASS
CO_2	44	0.35	15.4	$\frac{15.4}{27.3} = 0.564$
CH_4	16	0.20	3.2	$\frac{3.2}{27.3} = 0.117$
H_2	2	0.15	0.3	$\frac{0.3}{27.3} = 0.011$
N_2	28	0.30	8.4	$\frac{8.4}{27.3} = 0.308$
		1.00	27.3	1.000

The mass analysis is:

56.4% CO_2; 11.7% CH_4; 1.1% H_2; 30.8% N_2

11.15 Conversion of a mass analysis to a volumetric analysis

To convert from a mass analysis of a gas mixture to a volumetric analysis the reverse procedure to that given in Example 11.10 must be adopted. This is illustrated by Example 11.11.

Example 11.11

A gas analysis gave the following percentage analysis by mass

Carbon monoxide (CO) 40%; Oxygen (O_2) 15%;
Carbon dioxide (CO_2) 10%; Nitrogen (N_2) 35%.

Determine the percentage analysis by volume of the gas mixture.

Solution

CONSTITUENT	RELATIVE MOLECULAR MASS	FRACTION BY MASS	FRACTION BY MASS / RELATIVE MOLECULAR MASS	FRACTION BY VOLUME
CO	28	0.40	$\dfrac{0.40}{28} = 0.01429$	$\dfrac{0.01429}{0.03375} = 0.424$
O_2	32	0.15	$\dfrac{0.15}{32} = 0.00469$	$\dfrac{0.00469}{0.03375} = 0.139$
CO_2	44	0.10	$\dfrac{0.10}{44} = 0.00227$	$\dfrac{0.00227}{0.03375} = 0.067$
N_2	28	0.35	$\dfrac{0.35}{28} = 0.01250$	$\dfrac{0.01250}{0.03375} = 0.370$
Totals		1.00	0.03375	1.000

The percentage analysis by volume is:

42.4% CO; 13.9% O_2; 6.7% CO_2; 37% N_2

Problems

1. An air compressor takes in air at an atmospheric pressure of 100 kN/m² and a temperature of 14°C. If the compression ratio is 7:1 and the pressure after compression is 960 kN/m² what will be the final temperature?

2. Gas is contained in a cylinder, by means of a piston, at a pressure of 300 kN/m² and temperature of 15°C. If the gas temperature is raised to 400°C what additional force must be applied to the piston to prevent an increase in the volume of the gas. The cylinder diameter is 160 mm.

3. Air at a pressure of 105 kN/m² and temperature of 22°C is compressed to one-tenth of its initial volume such that its pressure is increased to 1.65 MN/m². Calculate the temperature of the air after compression.

4. A gas cylinder has a volume of 2.5 m³ and contains oxygen at a pressure of 2.8 MN/m² and temperature of 17°C. After using the cylinder the temperature is found to be 12°C and the pressure 2.0 MN/m². Calculate the volume of gas consumed at s.t.p.

5. The cylinder of an exhaust pump has a cross-sectional area of 0.016 m² and a stroke of 300 mm. It is connected to a vessel of 0.24 m³ capacity containing air at a pressure of 10^5 N/m². If the temperature is assumed to remain constant, determine the pressure of air in the vessel at the end of both the first and the second exhaust strokes.

6. A tyre has a volume of 0.053 m³ and contains air at a gauge pressure of 140 kN/m² and a temperature of 12°C. What volume of air at s.t.p. must be pumped into the tyre if its pressure is to be increased to 196 kN/m² (gauge)? Assume that the volume of the tyre increases by 8% and that the final temperature is 14°C.

7. Hydrogen is contained in a cylinder of volume 0.1 m³ at a pressure of 700 kN/m² and a temperature of 16°C. The cylinder is fitted with a safety valve which is designed to operate at a pressure of 780 kN/m². Determine:

(*a*) the temperature at which the safety valve will operate;
(*b*) the pressure of the gas in the cylinder after heating to 94°C and cooling to the original temperature;
(*c*) the volume of gas at s.t.p. lost from the cylinder during this process.

8. One kilogramme of a certain gas occupies a cylinder of volume 0.3 m³, the pressure of the gas being 136 kN/m² and the temperature 14°C. Calculate the value of the gas constant.

If a quantity of gas is now withdrawn from the cylinder such that the pressure drops to 40 kN/m², the temperature remaining constant, calculate the mass of gas which has been withdrawn.

9. A quantity of gas at a pressure of 570 kN/m² occupies a volume of 0.15 m³ at a temperature of 27°C. If the gas constant, R, is 188 J/kg K, calculate the mass of gas present.

If the pressure is now increased to 780 kN/m², with the volume remaining constant, what will be the final temperature of the gas?

10. A closed vessel having a volume of 1.2 m³ contains 11.2 kg of oxygen at a pressure of 700 kN/m² and temperature of 20°C. Determine the gas constant for oxygen and hence obtain the value of the specific heat capacity at constant pressure, c_p, given that the specific heat capacity at constant volume, c_v, is 670 J/kg K.

11. (*a*) State Boyle's law and Charles' law and show that they may be combined to produce the characteristic equation of state for a gas.

(b) Two kilogrammes of gas is pumped into an evacuated cylinder of internal capacity 0.2 m³. If the temperature of the gas when pumping is completed is 50°C, determine the pressure in the cylinder. Also determine the pressure in the cylinder when the temperature of the gas has fallen to 14°C. For the gas $R = 254$ J/kg K.

12. A gas cylinder contains 0.12 kg of hydrogen at a pressure of 600 kN/m² and a temperature of 15°C. Calculate the volume of the gas cylinder.

If 0.02 kg of the gas is consumed determine:

(a) the volume that this quantity of gas would occupy at s.t.p.
(b) the resulting pressure in the cylinder if the temperature falls to 2°C.
(R for hydrogen is 4.157 kJ/kg K.)

13. A gas cylinder 1.2 m long and 0.25 m internal diameter contains carbon dioxide (CO_2) gas at a pressure of 3 MN/m² and temperature of 17°C. Calculate the mass of gas in the cylinder.
($R_0 = 8314$ J/kg mol K.)

14. A storage tank having a volume of 3.6 m³ contains a gas at a pressure of 1.9 MN/m² and temperature of 12°C. What mass of gas must be removed from the tank to reduce the pressure to 1 MN/m² with the temperature at 8°C? ($R = 287$ J/kg K.)

15. The initial volume of 0.3 kg of gas was 0.25 m³ at a temperature of 16°C and pressure of 90 kN/m². Determine:

(a) the relative molecular mass of the gas;
(b) the specific heat capacity at constant pressure; c_p.
($R_0 = 8314$ J/kg mol K; $c_p = 1.41 c_v$).

16. 1.8 kg of oxygen are contained in a cylinder, having a volume of 2.4 m³, at a temperature of 12°C. Calculate the pressure of the gas. ($R_0 = 8314$ J/kg mol K.)

17. A sealed chamber has dimensions of 6 m × 4.8 m × 3.6 m. If the pressure in the chamber is 100 kN/m² when the temperature is 4°C, what will be the rise in pressure if the temperature rises to 20°C, assuming no loss of air. What mass of air will escape when the chamber door is opened? Take the atmospheric pressure as 100 kN/m² and the gas constant as 288 J/kg K.

18. A sphere having an internal diameter of 1.5 m contains the following gas mixture:

Carbon dioxide (CO_2),	0.5 kg;
Nitrogen (N_2),	0.4 kg;
Hydrogen (H_2),	1.6 kg.

Determine the resultant pressure in the sphere when the temperature is 90°C. ($R_0 = 8314$ J/kg mol K.)

19. A vessel of volume 0.6 m^3 contains 0.7 kg of oxygen and 1.7 kg of an unknown gas at a combined pressure of 300 kN/m^2 and a temperature of 48°C. Determine the partial pressures of each gas and the characteristic gas constant for the unknown gas. (For oxygen, $R = 260$ J/kg K.)

20. A vessel having a volume of 0.1 m^3 contains a mixture of three gases A, B, and C. The masses of each gas are:

A, 0.3 kg; B, 0.7 kg; C, 1.05 kg.

If the relative molecular masses of A, B, and C are 24, 29, and 18 respectively, calculate the pressure in the vessel when the temperature is 60°C. ($R_0 = 8314$ J/kg mol K.)

21. A vessel of 0.65 m^3 capacity contains air at a pressure of 360 kN/m^2 and a temperature of 12°C. Determine the masses of oxygen and nitrogen and their partial pressures.

If hydrogen at the same temperature is added, increasing the pressure in the vessel to 400 kN/m^2, the temperature remaining the same, what would be the mass and partial pressure of the hydrogen? (Assume: air contains 23.3% oxygen by mass,

R for air   = 288 J/kg K,
R for oxygen = 260 J/kg K,
R for hydrogen = 4157 J/kg K.)

22. A sample of dry exhaust gas from an engine consists of 2.5 m^3 nitrogen (N_2), 0.035 m^3 carbon monoxide (CO), and 0.2 m^3 carbon dioxide (CO_2). Determine the percentage analysis by mass of the exhaust gas.

23. The volumetric analysis of a sample of coal gas is: hydrogen (H_2), 46%; carbon monoxide (CO), 15%; methane (CH_4), 35%; nitrogen (N_2), 4%. Calculate the percentage analysis of the gas by mass.

24. A boiler is supplied with gas having the following volumetric analysis:

Carbon monoxide (CO), 16%;   Hydrogen (H_2), 34%;
Nitrogen (N_2),   27%;   Methane (CH_4), 23%.

Calculate the percentage analysis by mass of this gas.

25. A gas has the following analysis by mass:

$\quad$ Carbon dioxide (CO_2), 16%;
$\quad$ Oxygen (O_2), 47%;
$\quad$ Nitrogen (N_2), 37%.

What is the percentage analysis by volume?

26. An analysis of a gas gave the following composition by mass:

$\quad$ Carbon monoxide (CO), 0.10 kg;
$\quad$ Nitrogen (N_2), 0.07 kg;
$\quad$ Hydrogen (H_2), 0.09 kg;
$\quad$ Oxygen (O_2), 0.05 kg.

Determine the percentage analysis by volume.

12

Combustion

12.1 Combustion

Combustion is a chemical action in which fuel elements or compounds are rapidly oxidised and heat energy is released. Combustion can only occur if:

(a) there is a continuous supply of oxygen;
(b) there is a sufficiently high temperature to start the combustion process.

The oxygen is usually obtained from air. This contains 23.3% oxygen by mass and 21% by volume. The other major constituent of air is nitrogen and this is chemically inactive at most combustion temperatures.

Any combustion process can be represented by a chemical equation in which the total mass of any element must be the same on either side of the equation.

12.2 Fuels

Fuels can be solid, liquid, or gaseous. Examples of these are given below:

CATEGORY	EXAMPLES
Solid	Coal, wood, anthracite
Liquid	Petrol, paraffin, oil (diesel)
Gaseous	Natural gas (methane), town gas (commercially prepared), hydrogen

Atomic fuels are not dealt with here as no oxidation occurs during the release of heat energy.

The major combustible constituents of most fuels are carbon or hydrogen, or both, hence the term 'hydrocarbons'. Solid fuels usually possess very little hydrogen and are frequently nothing more than solid carbon. In addition to the combustible elements of a fuel there is always some incombustible material present. This remains as ash and is often detrimental to the combustion of a fuel as it can impede the air flow which is vital for satisfactory combustion. In fact, the excessive presence of these incombustibles in low grade coals and peat often means that the substance is completely useless as a fuel, in spite of the fact that a great deal of combustible material is present.

12.3 Atoms and molecules

An atom is the smallest particle retaining the characteristics of an element. Although each atom has a mass, it is very small and in most instances the mass of an atom is quoted relative to that of another atom. This gives rise to a table of relative atomic masses and the following table gives the values for those elements which we shall encounter in combustion equations.

ELEMENT	CHEMICAL SYMBOL	RELATIVE ATOMIC MASS
Hydrogen	H	1
Carbon	C	12
Nitrogen	N	14
Oxygen	O	16
Sulphur	S	32

Although the atom is the smallest quantity of a single element it does not follow that it can exist on its own. In many instances two or more

ELEMENT	NO. OF ATOMS	MOLECULAR SYMBOL	RELATIVE MOLECULAR MASS
Hydrogen	2	H_2	2
Carbon	1	C	12
Nitrogen	2	N_2	28
Oxygen	2	O_2	32
Sulphur	1	S	32

atoms of an element combine to form a stable particle, known as a 'molecule'. Thus, we have relative molecular masses, as given above:
 A compound is a chemical combination of two or more elements to produce a new substance. The properties of the compound are completely different from those of its constituent elements, e.g. salt (sodium chloride) is a compound of sodium, a soft whitish metal, and chlorine, a greenish poisonous gas having a pungent odour. The molecular mass of a compound is the sum of the atomic masses of its constituent atoms.

12.4 Combustion of elements

HYDROGEN

$$\text{Hydrogen} + \text{oxygen} \rightarrow \text{water (steam)}$$
$$2H_2 + O_2 \rightarrow 2H_2O$$

This equation cannot be written as

$$H_2 + O \rightarrow H_2O$$

since this implies a single atom of oxygen, whereas it exists in the stable molecular form.

By volume: 2 vol. H_2 + 1 vol. O_2 → 2 vol. H_2O

By mass: 4 kg H_2 + 32 kg O_2 → 36 kg H_2O

or   1 kg H_2 + 8 kg O_2 → 9 kg H_2O

Thus, 1 kg of hydrogen requires 8 kg of oxygen for complete combustion and as a result 9 kg of water (steam) is produced.

CARBON

$$\text{Carbon} + \text{oxygen} \rightarrow \text{carbon dioxide}$$
$$C^* + O_2 \rightarrow CO_2$$

By volume: 0* + 1 vol. O_2 → 1 vol. CO_2

By mass: 12 kg C + 32 kg O_2 → 44 kg CO_2

or   1 kg C + $\tfrac{32}{12}$ kg O_2 → $\tfrac{44}{12}$ kg CO_2

Thus 1 kg of carbon requires $\tfrac{32}{12}$ kg of oxygen for complete combustion and as a result $\tfrac{44}{12}$ kg of carbon dioxide is produced.
 If insufficient oxygen is present for complete combustion of the car-

bon then carbon monoxide (CO) will be formed as well as carbon dioxide.

*Note: As carbon is a solid fuel it has a negligible volume compared with a gas.

SULPHUR

$$\text{Sulphur} + \text{oxygen} \rightarrow \text{sulphur dioxide}$$
$$S \ + \ O_2 \ \rightarrow SO_2$$

By volume: $\quad 0 + 1 \text{ vol. } O_2 \rightarrow 1 \text{ vol. } SO_2$

By mass: $\quad 32 \text{ kg S} + 32 \text{ kg } O_2 \rightarrow 64 \text{ kg } SO_2$

or $\quad 1 \text{ kg S} + 1 \text{ kg } O_2 \rightarrow 2 \text{ kg } SO_2$

Thus, 1 kg of sulphur requires 1 kg of oxygen for complete combustion and as a result 2 kg of sulphur dioxide is formed.

Note: Sulphur, like carbon, is a solid and therefore has a negligible volume compared with a gas.

12.5 Minimum air required for combustion

It has been mentioned in Section 12.1 that air contains 23.3% oxygen by mass and 21% oxygen by volume, with the remainder being mainly nitrogen. As the oxygen for a combustion process, whether it be the combustion of a solid fuel in a boiler or liquid fuel in an engine, is supplied from the atmosphere it is necessary to know the theoretical minimum quantity of air required for combustion. Thus, from the above mass and volume analysis of air we obtain:

Mass of air required for combustion

$$= \frac{100}{23.3} \times \text{mass of oxygen}$$

$$= 4.29 \times \text{mass of oxygen}$$

Volume of air required for combustion

$$= \frac{100}{21} \times \text{vol. of oxygen}$$

$$= 4.76 \times \text{vol. of oxygen}$$

In practice, however, complete combustion would not be likely to occur if such quantities of air were supplied. This is because it is virtually impossible to achieve the distribution of oxygen necessary to produce complete combustion. For example, in an internal combustion engine the mixing process inside the cylinder is not so perfect as to produce combustion of all the carbon present in the fuel to carbon dioxide in the very short time available. Instead, a mixture of carbon dioxide and carbon monoxide (CO), which is the incomplete combustion product of carbon, is produced. Thus, it is always necessary to supply excess air if complete combustion of the fuel is to be possible and even then, if the process is very rapid and in an enclosed space, complete combustion will not necessarily occur.

When the quantity of air supplied is just sufficient for the complete combustion of the fuel it is known as the chemically correct, or stoichiometric, air to fuel ratio.

12.6 Excess air supply

In Section 12.5 it was stated that if the theoretical minimum quantity of air were supplied to a combustion process it was unlikely that complete combustion would occur. Thus, it is usual to supply an excess quantity of air to ensure that there is a plentiful supply of oxygen available and so enhance the chances of achieving complete combustion of the fuel. The actual air/fuel ratio depends upon many requirements, e.g. rate of combustion required, chilling due to excess air, heat losses in exhausting gases, etc. and will not be dealt with further at this stage.

The correct air/fuel ratio is expressed in terms of mass for solid and liquid fuels and by volume for gaseous fuels.

$$\text{Percentage excess air} = \frac{\text{actual air/fuel ratio} - \text{chemically correct air/fuel ratio}}{\text{chemically correct air/fuel ratio}} \times 100\%$$

$$= \frac{\text{total air supplied} - \text{minimum air required}}{\text{minimum air required}} \times 100\%$$

When referring to a petrol engine an air/fuel mixture containing an excess of air is known as a weak mixture while one which is deficient in air is known as a rich mixture. Thus, when starting a cold engine it is

Combustion

usually to provide a rich mixture by using the choke, so restricting the air supply.

$$\text{Mixture strength} = \frac{\text{chemically correct air/fuel ratio}}{\text{actual air/fuel ratio}} \times 100\%$$

Example 12.1

A fuel has the following analysis by mass: carbon 68%, hydrogen 21%, oxygen 10%, incombustible impurities 1%. Determine the mass of air required for the complete combustion of 1 kg of the fuel.

Solution

Combustion of carbon:

From Section 12.4, 1 kg of C requires $\frac{32}{12}$ kg of O_2 for complete combustion. Now 1 kg of the fuel contains 0.68 of carbon, therefore 0.68 kg of C require $0.68 \times \frac{32}{12} = 1.81$ kg of O_2 for complete combustion

Combustion of hydrogen:

From Section 12.4, 1 kg of H_2 requires 8 kg of O_2 for complete combustion. But 1 kg of the fuel contains 0.21 kg of hydrogen, therefore 0.21 kg of H_2 require $0.21 \times 8 = 1.68$ kg of O_2 for complete combustion.

Thus, the quantity of oxygen required for the complete combustion of the hydrogen and carbon in the fuel is $1.81 + 1.68 = 3.49$ kg.

But 1 kg of the fuel contains 0.10 kg of oxygen and this is assumed to be available for combustion.

∴ Oxygen required from the air $= 3.49 - 0.10 = 3.39$ kg

But air consists of 23.3% O_2 by mass.

∴ Mass of air required/kg of fuel $= \dfrac{3.39 \times 100}{23.3}$

$= 14.55$ kg

This type of problem can be solved conveniently by a tabular method, as shown below.

CONSTITUENT	MASS/kg OF FUEL	O_2 REQUIRED/kg	O_2 REQUIRED	
C	0.68	$\frac{32}{12}$	$0.68 \times \frac{32}{12} =$	1.81 kg
H_2	0.21	8	$0.21 \times 8 =$	1.68 kg
O_2	0.10	—		-0.10 kg
Impurities	0.01	—		—
	1.0			3.39 kg

Example 12.2

(a) Determine the minimum quantity of air required for the complete combustion of 1 kg of hexane, C_6H_{14}.

(b) If the actual air/fuel ratio supplied is 14:1 calculate the mixture strength.

Solution

(a) Method 1—by combustion equation.

$$C_6H_{14} + 9.5O_2 \rightarrow 6CO_2 + 7H_2O$$

The combustion equation for the fuel is given above and is obtained by balancing the atoms on either side of the equation. As there are 6 carbon atoms in the fuel then there must be $6CO_2$ produced. Similarly, as there are 14 hydrogen atoms in the fuel then $7H_2O$ must be produced. Thus the oxygen supplied must balance the output, i.e.

$$\text{Oxygen supplied} = 6O_2 + 7O$$

As there are two atoms in an oxygen molecule it follows that the 7 atomic units of oxygen are equivalent to 3.5 molecular units. Thus,

$$\text{Oxygen supplied} = 6O_2 + 3.5O_2$$
$$= 9.5O_2$$

Using the atomic masses of the element gives $(6 \times 12 + 14 \times 1)$ kg C_6H_{14} requires (9.5×32) kg O_2 for complete combustion or 86 kg of C_6H_{14} requires 304 kg O_2 for complete combustion.

$$\therefore 1 \text{ kg } C_6H_{14} \text{ requires } \tfrac{304}{86} = 3.53 \text{ kg } O_2$$

for complete combustion

$$\therefore \text{Mass of air required/kg of hexane} = \frac{3.53}{0.233}$$
$$= 15.15 \text{ kg}$$

Method 2—by constituent analysis

$$\text{Molecular mass of hexane, } C_6H_{14} = (6 \times 12) + (14 \times 1)$$
$$= 72 + 14$$
$$= 86$$

Thus the fuel contains

$\frac{72}{86} \times 100 = 83.7\%$ carbon by mass

and $\frac{14}{86} \times 100 = 16.3\%$ hydrogen by mass

CONSTITUENT	MASS/kg OF FUEL	O_2 REQUIRED/kg	O_2 REQUIRED
C	0.837	$\frac{32}{12}$	$0.837 \times \frac{32}{12} = 2.23$ kg
H_2	0.163	8	$0.163 \times 8 = 1.30$ kg
	1.00		3.53 kg

Then, Minimum air required $= \dfrac{3.53}{0.233} = 15.15$ kg

The mass of air required to completely combust 1 kg of hexane, C_6H_{14}, is 15.15 kg.

(b) Mixture strength $= \dfrac{\text{correct air/fuel ratio}}{\text{actual air/fuel ratio}} \times 100\%$

$= \dfrac{15.15}{14} \times 100$

$= 108.2\%$

i.e. the mixture is 8.2% rich.

12.7 Products of combustion

The products of a combustion process are mainly gaseous and consist of carbon dioxide, water vapour, sulphur dioxide, oxygen, and nitrogen. However, if insufficient oxygen is supplied for complete combustion, or if the temperature is very high, then carbon monoxide will be present in the exhaust gases as well as carbon dioxide.

An analysis of the exhaust gases enables the efficiency of the combustion process to be determined, but such an analysis, if it is to be of any value to the engineer, must be reasonably accurate. In most instances the sample to be analysed is cooled, with the result that any steam present is condensed. Such an analysis gives the composition of the dry products. If the steam is included in the exhaust analysis then it is known as a wet analysis.

As the combustion products are gaseous any exhaust analysis is usually on a volumetric basis and is frequently obtained by means of an Orsat apparatus, Fig. 12.1.

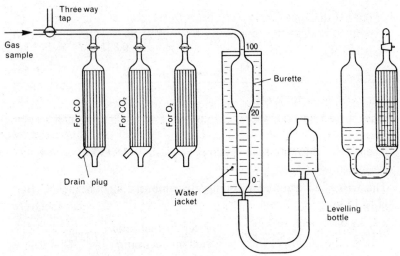

12.1 Orsat apparatus

With this apparatus the analysis is by chemical means and the necessary solutions are contained in three separate vessels which are commonly connected to a burette, calibrated from 0 to 100. At the left-hand side of this connecting tube is a three-way tap by means of which the apparatus can be connected to the gas sample to be analysed, to the atmosphere, or used to seal the apparatus. The burette is surrounded by a water jacket to maintain a uniform temperature. A levelling bottle or aspirator is connected to the burette and this is used to manipulate the gas sample and chemical solutions.

The three vessels are filled with solutions to absorb the carbon dioxide, oxygen, and carbon monoxide respectively. The usual solutions used are as follows:

For carbon dioxide: a solution of caustic soda or caustic potash.
For oxygen: a pyrogallic acid solution.
For carbon monoxide: a cuprous chloride solution.

It is imperative that the gases are absorbed in the above order since carbon dioxide is also soluble in the pyrogallic acid and both carbon dioxide and oxygen are soluble in the cuprous chloride solution.

With the three-way tap open to the atmosphere the air in the connecting tube is expelled by raising the levelling bottle. The levelling bottle is now lowered and by opening each tap in turn the levels of the chemical solutions can be adjusted to just below the level of the tap. The three-way tap is again opened to the atmosphere and the air in the burette expelled by raising the levelling bottle. The three-way tap is now connected to the gas sample to be analysed. The levels of the burette and levelling bottle are set to zero and the three-way tap sealed. The tap to the first absorption vessel (CO_2) is now opened and the levelling bottle raised to force the gas through the chemical solution. This is repeated until the same burette reading is obtained for two successive readings, with the same level in the burette and levelling bottle. The burette reading will give the percentage CO_2 in the gas. The process is repeated for the absorption of the oxygen and carbon monoxide, the difference between the initial and final burette readings giving the percentage of the respective gas. The remaining gas in the burette is assumed to be nitrogen.

To ensure that a dry gas is being analysed the sample is sometimes passed through a drying agent, such as calcium chloride contained in a U-tube prior to the three-way valve. Even without this a true analysis of the dry products is obtained. It should be noticed that any sulphur dioxide present in the gas sample is not covered by this analysis.

The analysis of the exhaust gases of a solid fuel may not present a complete breakdown since any ash and unburnt carbon will not be present in the exhaust gases. Hence, the residual solid particles must be analysed as well.

Examples illustrating analysis by mass and volume and the conversion of one analysis to another now follow.

Example 12.3

The percentage composition by mass of anthracite coal used to fire a boiler is:

Carbon 82%; Hydrogen 4.5%; Oxygen 8%
Nitrogen 2%; Sulphur 1.5%; Ash 2%

Determine the mass of air required for the complete combustion of 1 kg of the coal and the percentage analysis by mass of the products of combustion.

Solution

CONSTITUENT	MASS/kg OF COAL	O_2 REQUIRED (kg)	COMBUSTION PRODUCT
C	0.82	$0.82 \times \frac{32}{12} = 2.19$	$0.82 + 2.19 = 3.01$ kg CO_2
H_2	0.045	$0.045 \times 8 = 0.36$	$0.045 + 0.36 = 0.405$ kg H_2O
O_2	0.08	-0.08	—
N_2	0.02		0.02 kg N_2
S	0.015	$0.015 \times 1 = 0.015$	$0.015 + 0.015 = 0.03$ kg SO_2
Ash	0.02	—	
	1.00	2.485	

Mass of O_2 required/kg of coal = 2.485 kg

Since air consists of 23.3% O_2 by mass.

$$\text{Mass of air required/kg of coal} = \frac{2.485}{0.233}$$

$$= 10.66 \text{ kg}$$

Mass of nitrogen present in air supplied = $10.66 - 2.485$

$$= 8.175 \text{ kg}$$

Therefore,

Total mass of nitrogen in the exhaust gases = $8.175 + 0.02$

$$= 8.195 \text{ kg}$$

The analysis of the combustion products is then:

PRODUCT	MASS(kg)	% BY MASS
CO_2	3.01	$\frac{3.01}{11.64} \times 100 = 25.86$
H_2O	0.405	$\frac{0.405}{11.64} \times 100 = 3.48$
N_2	8.195	$\frac{8.195}{11.64} \times 100 = 70.4$
SO_2	0.03	$\frac{0.03}{11.64} \times 100 = 0.26$
	11.64	100.0

Combustion 225

The analysis by mass of the products of combustion is

25.86% CO_2, 3.48% H_2O, 70.40% N_2, 0.26% SO_2

Example 12.4

The volumetric analysis of a gas is:

22% H_2, 31% CO, 16% CH_4, 4% CO_2, 27% N_2

Determine:

(a) the volume of air required for the complete combustion of 1000 l (litre) of the gas;
(b) the analysis by volume of the exhaust products.

Solution

(a)
Combustion of hydrogen:

$$2H_2 + O_2 \rightarrow 2H_2O$$

2 vol. H_2 + 1 vol. $O_2 \rightarrow$ 2 vol. H_2O

∴ For the 220 l of H_2 contained in 1000 l of the gas

220 l H_2 + 110 l $O_2 \rightarrow$ 220 l H_2O

Combustion of carbon monoxide:

$$2CO + O_2 \rightarrow 2CO_2$$

2 vol. CO + 1 vol. $O_2 \rightarrow$ 2 vol. CO_2

∴ For the 310 l of CO contained in 1000 l of the gas

310 l CO + 155 l $O_2 \rightarrow$ 310 l CO_2

Combustion of methane:

$$CH_4 + 2O_2 \rightarrow CO_2 + 2H_2O$$

1 vol. CH_4 + 2 vol. $O_2 \rightarrow$ 1 vol. CO_2 + 2 vol. H_2O

∴ For the 160 l of CH_4 contained in 1000 l of the gas

160 l CH_4 + 320 l $O_2 \rightarrow$ 160 l CO_2 + 320 l H_2O

Thus

Total volume of oxygen required = 110 + 155 + 320
= 585 l

Since air contains 21% oxygen by volume,

$$\text{Volume of air required} = \frac{595}{0.21} = 2786 \text{ l}$$

(b)   Volume of nitrogen supplied $= 2786 - 585$
$$= 2201 \text{ l}$$

∴ Total volume of N_2 in exhaust gas $= N_2$ supplied $+ N_2$ in gas
$$= 2201 + 270$$
$$= 2471 \text{ l}$$

Total CO_2 in exhaust $= CO_2$ produced $+ CO_2$ in gas
$$= (310 + 160) + 40$$
$$= 510 \text{ l}$$

The analysis of the gas is then:

PRODUCT	VOLUME (litres)	% VOLUME
CO_2	510	$\frac{510}{3521} \times 100 = 14.48$
H_2O	$220 + 320 = 540$	$\frac{540}{3521} \times 100 = 15.34$
N_2	2471	$\frac{2471}{3521} \times 100 = 70.18$
	3521	100.0

A mass analysis could be converted to a volumetric analysis, or vice versa by the method given in Chapter 11, Examples 11.8 and 11.9.

Example 12.5

If the air supplied in Example 12.3 is 35% in excess of that required for complete combustion calculate the analysis by mass and by volume of the dry combustion products.

Solution

From Example 12.3 the quantity of air required for the complete combustion of 1 kg of coal $= 10.66$ kg

∴ Excess air supplied $= 0.35 \times 10.66 = 3.733$ kg

This excess air contains

$$0.233 \times 3.733 = 0.870 \text{ kg } O_2$$

and $\quad 0.767 \times 3.733 = 2.863 \text{ kg } N_2$

These quantities will be additional to the combustion products given in Example 12.3, but for a dry analysis the H_2O is not included.

The analysis by mass is then converted into a volumetric analysis as for Example 11.10.

PRODUCT	MASS IN PRODUCTS (kg)	% BY MASS	% BY MASS / MOL. MASS	% BY VOL.
CO_2	3.01	20.11	$\dfrac{20.11}{44} = 0.457$	13.93
N_2	8.195 + 2.863 = 11.058	73.88	$\dfrac{73.88}{28} = 2.638$	80.43
SO_2	0.03	0.20	$\dfrac{0.20}{64} = 0.003$	0.09
O_2	0.87	5.81	$\dfrac{5.81}{32} = 0.182$	5.55
	14.968	100	3.280	100

The analysis of the dry combustion products is:

By mass:

20.11% CO_2; 73.88% N_2; 0.20% SO_2; 5.81% O_2

By volume:

13.93% CO_2; 80.43% N_2; 0.09% SO_2; 5.55% O_2

12.8 Calorific value of a fuel

The quantity of heat energy liberated during the combustion of unit mass or unit volume of a fuel is known as its calorific value (C.V.).

The higher calorific value (H.C.V.) of a fuel is the energy output from the complete combustion of unit quantity of the fuel when the combustion products are cooled to the original temperature of the fuel.

228   Mechanical engineering science

The lower calorific value (L.C.V.) of a fuel is obtained by subtracting an allowance for the non-condensation of the steam from the higher calorific value. This arises because it is usually impracticable to cool the combustion products down to the original temperature and hence the steam does not release the energy it possesses. Thus,

$$\text{L.C.V.} = \text{H.C.V.} - m_c \times h_{fg} \qquad (12.1)$$

where m_c is the mass of condensate (steam) produced per unit mass or volume of the fuel, and h_{fg} is the latent heat of vaporisation of water, usually taken at a temperature of 25°C. At 25°C, $h_{fg} = 2442$ kJ/kg.

Example 12.6

A fuel oil has the following mass analysis:

Carbon 79%, Hydrogen 10%, Oxygen 8%
Sulphur 2.5%, Impurities 0.5%

Determine the higher and lower calorific values taking the condensate allowance as 2442 kJ/kg of vapour. Assume the calorific values to be:

Carbon   33 750 kJ/kg
Hydrogen   161 000 kJ/kg
Sulphur   9 070 kJ/kg

Solution

For the combustion of 1 kg of the fuel the H.C.V. is obtained from the sum of the energies released by the individual constituents

Energy from Carbon   = 0.79 × 33 750 = 26 662 kJ
Energy from Hydrogen = 0.10 × 161 000 = 16 100 kJ
Energy from Sulphur = 0.025 × 9 070 =   227 kJ

∴ For fuel, H.C.V. = 42 989 kJ

From Section 12.4

$$1 \text{ kg } H_2 + 8 \text{ kg } O_2 \rightarrow 9 \text{ kg } H_2O$$

Thus from the combustion of 0.10 kg H_2 the mass of condensate formed is 0.9 kg. Then,

L.C.V. = H.C.V. − condensate allowance
      = 42 989 − 2442 × 0.9
      = 40 791 J/g

The H.C.V. of the fuel is 42 989 kJ/kg and its L.C.V. is 40 791 kJ/kg.

Example 12.7

In an experiment to determine the calorific value of a coal sample it was found that on burning 1.4 g of the coal the temperature of 1.8 kg of water contained in a bomb calorimeter apparatus having a water equivalent of 620 g rose from 16.1°C to 17.5°C. If the temperature correction for cooling losses is 0.1°C calculate the calorific value of the coal. (Specific heat capacity of water = 4.18 J/kg K.)

Solution

If no heat losses had occurred during the experiment the final temperature would have been 17.6°C. Thus,

$$\text{Corrected temperature rise} = 17.6 - 16.1$$
$$= 1.5°C$$

$$\text{Total water equivalent of apparatus} = 1.8 \times 1000 + 620$$
$$= 2420 \text{ g}$$

$$\therefore \text{Heat transferred to apparatus and water} = 2420 \times 1.5 \times 4.18$$
$$= 15\,173 \text{ J}$$

$$\therefore \text{Calorific value of fuel} = \frac{15\,173}{1.4} \frac{[\text{J}]}{[\text{g}]}$$
$$= 10\,840 \text{ J/g}$$
$$= 10.84 \text{ kJ/g}$$

The calorific value of the coal is 10.84 kJ/g.

12.9 Determination of the calorific value of a fuel

The experimental determination of the calorific value of a fuel is carried out in a special calorimeter. In the case of a solid, and some liquid fuels, the bomb calorimeter is used; while for gases, and some liquid fuels, the gas calorimeter is used.

(*a*) THE BOMB CALORIMETER

With this calorimeter, shown in Fig. 12.2, a small quantity of fuel is contained in a crucible within a stainless steel vessel (bomb). At the top of this bomb is a one-way oxygen admission valve and a pressure release valve. The bomb is surrounded by a water container which, in turn, is surrounded by an air-jacket to provide heat insulation. A stirrer and thermometer which is usually of the fixed range or Beckman type, are fitted as shown.

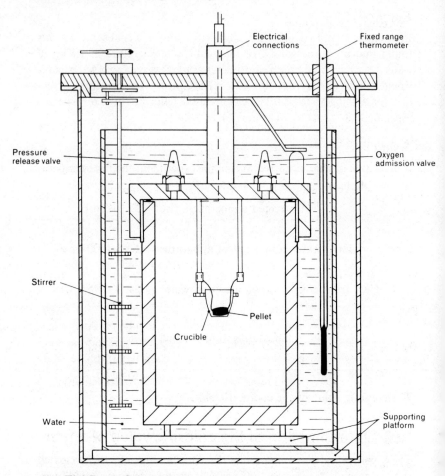

12.2 The Bomb Calorimeter

The quantity of fuel used is estimated such that a small temperature rise of 2–3 K is produced. In the case of a solid fuel it is usually crushed, passed through a fine sieve, and ground into a small pellet. The pellet is ignited by fusing a small platinum or nichrome wire which is in contact with it. This wire forms part of an electrical circuit which includes a firing switch. The procedure is to add a known mass of the fuel to the crucible and arrange the fuse to contact the fuel. It is worth noting that by using a special press the pellet can be produced with the fuse passing through it. The crucible is positioned inside the bomb and a small quantity of water is poured into the bottom of the bomb. The water absorbs

Combustion

the combustion vapours and ensures that any steam is condensed. The bomb is now sealed and slowly pressurised to 23–25 atmospheres with oxygen. The valves are then sealed and the bomb placed within the water vessel. A known quantity of water is now added to the water vessel so that the bomb is immersed. The cover and stirrer are fitted and the thermometer immersed in the water to a suitable depth. The electrical connections are made and the rheostat in the circuit adjusted to produce a suitable current.

Stirring is commenced and the temperature recorded at minute intervals for a period of five minutes. At the end of this time the charge is fired and the temperature recorded at 10 second intervals until the readings start to fall when they can be taken every minute for a further 5 minutes. At the end of this time the bomb is removed, the pressure release valve opened, and the crucible inspected to ensure that the fuel has been completely fired.

When determining the calorific value of a liquid fuel it is contained in a gelatin capsule.

Calculation of result

The measured temperature rise must be corrected for various losses, the largest being the cooling losses. This can be determined by graphical means but British Standards recommend the adoption of the Regnault–Pfaundler cooling correction. This is:

$$\text{Correction} = nv + \left(\frac{v_1 - v}{t_1 - t}\right) \left\{ \sum_{1}^{n=1} (t) + \tfrac{1}{2}(t_0 + t_n) - nt \right\}$$

where n = no. of minutes between firing and first reading after temperature begins to fall,

v = rate of temperature fall during the pre-combustion period,

v_1 = rate of temperature fall after the maximum temperature is reached,

t = average temperature during pre-combustion period,

t_1 = average temperature in period after maximum temperature,

$\sum_{1}^{n-1}(t)$ = sum of temperature between firing and commencement of cooling,

$\tfrac{1}{2}(t_0 + t_n)$ = mean of firing temperature (t_0) and the first temperature after the rate of change of temperature becomes constant (t_n).

Example 12.8

The following results were obtained during a test with a bomb calorimeter. The mass of anthracite burned was 0.775 g and the total water equivalent of the apparatus 2.720 kg. Determine the calorific value of anthracite. (Specific heat capacity of water = 4.18 kJ/kg K.)

PRE-FIRING PERIOD		HEATING PERIOD		COOLING PERIOD	
TIME (min)	TEMP. (°C)	TIME (min)	TEMP. (°C)	TIME (min)	TEMP. (°C)
0	15.526	t_1 6	17.414	t_n 10	17.866
1	15.528	t_2 7	17.862	11	17.863
2	15.530	t_3 8	17.871	12	17.860
3	15.532	t_4 9	17.875	13	17.856
4	15.534			14	17.853
t_0 5	15.536			15	17.850

Solution

$t = 15.531°C$

$v = \dfrac{15.526 - 15.536}{5}$

$= -0.002°C/\text{min}$

(negative sign indicates rising temperature)

$\sum_{1}^{n-1}(t) = 71.022$

$n = 5$

$\tfrac{1}{2}(t_0 + t_n)$

$= \dfrac{15.536 + 17.866}{2}$

$= 16.701°C$

$t_1 = 17.858°C$

$v_1 = \dfrac{17.866 - 17.850}{5}$

$= 0.0032°C/\text{min}$

Then,

$$\text{Cooling correction} = -0.002 \times 5 + \left(\dfrac{0.0032 + 0.002}{17.858 - 15.531}\right)$$
$$\times (71.022 + 16.701 - 5 \times 15.531)$$
$$= +0.012°C$$

Uncorrected temperature rise $= t_n - t_0$

$= 17.866 - 15.536$

$= 2.33°C$

∴ Corrected temperature rise $= 2.33 + 0.012$

$= 2.342°C$

Combustion   233

Energy released by fuel = 2.72 [kg] × 2.342 [°C] × 4.18 [kJ/kg K]
= 26.64 kJ

But this energy was obtained from 0.775 g of anthracite

∴ Calorific value of anthracite = $\dfrac{26.64}{0.775 \times 10^{-3}} \dfrac{[kJ]}{[kg]}$

= 34 374 kJ/kg

Note: The water equivalent of the calorimeter is determined by burning a known mass of a fuel of known calorific value, such as benzoic acid, in the bomb.

(b) THE GAS CALORIMETER (BOY'S CALORIMETER)

This is used for the determination of the calorific value of a gaseous fuel and a diagram is given in Fig. 12.3.

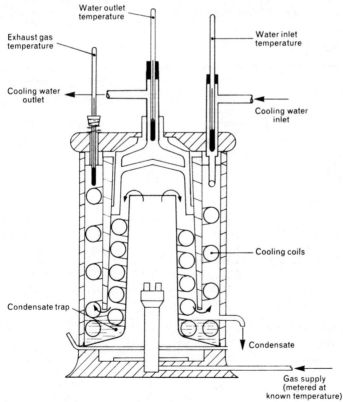

12.3 Boy's Gas Calorimeter

234 Mechanical engineering science

A continuous supply of the gas, at a constant pressure is metered and burned in an excess air supply. The exhaust gases pass over cooling coils and are exhausted at the base, where their temperature is measured. The mass of cooling water used in a certain time is measured, as are the inlet and outlet temperatures of this water. The water vapour in the exhaust gases will condense as the gases pass over the cooling coils and this condensate is collected in the condensate trap, from which it passes into a beaker for its mass to be determined.

The calorific value of a gas is quoted per unit volume, and as the volume depends upon pressure and temperature it is necessary to refer the volume to standard temperature and pressure conditions (s.t.p.) which are: 101.3 kN/m^2 or 760 mm Hg, and $0°C$.

The determination of the calorific value from the results is illustrated in Example 12.9.

The calorific value of a fuel oil can be determined with this calorimeter by fitting a suitable burner and supplying the oil under pressure from an external container.

Typical calorific values are given in Table 4.

Table 4. Typical calorific values of fuels

FUEL	CALORIFIC VALUE (kJ/kg)		CALORIFIC VALUE (kJ/m^3)	
	HIGHER	LOWER	HIGHER	LOWER
Anthracite	34 580	33 910		
Coal	33 500	32 400		
Coke	30 700	30 450		
Petrol	46 900	43 700		
Paraffin	46 500			
Diesel oil	45 350	42 650		
Hydrogen	160 000	120 000	11 930	10 060
Carbon to CO	10 100			
Carbon to CO_2	33 900			
CO to CO_2	10 200	10 200	11 850	
Methane	55 500	49 950	37 730	
Sulphur to SO_2	9 140			
Coal gas			20 150	17 950
Producer gas			6 070	6 040
Natural gas			36 400	32 800

Example 12.9

In a test to determine the calorific value of a gas, using a gas calorimeter, the following results were obtained:

Combustion

Volume of gas consumed	$= 0.0126 \text{ m}^3$
Gas temperature	$= 288 \text{ K}$
Gas manometer reading	$= 97 \text{ mm H}_2\text{O}$
Barometric height	$= 756 \text{ mm}$
Mass of cooling water used in same time as gas	$= 3.7 \text{ kg}$
Mass of condensate collected in same time as gas	$= 1.6 \text{ g}$
Cooling water inlet temperature	$= 285 \text{ K}$
Cooling water outlet temperature	$= 294 \text{ K}$

Determine the higher and lower calorific values of the gas.
(Specific heat capacity of water $= 4.18$ kJ/kg K.)

Solution

Assuming the relative density of mercury to be 13.6 then the gas pressure is equivalent to

$$756 + \frac{97}{13.6} = 763.13 \text{ mm mercury}$$

As the calorific value is quoted per unit volume it is necessary to calculate the volume at standard temperature and pressure conditions, i.e. 273 K and 760 mm mercury.

Applying $\qquad \dfrac{P_1 V_1}{T_1} = \dfrac{P_s V_s}{T_s}$

where $P_1 = 763.13$ mm Hg, $\quad P_s = 760$ mm Hg

$V_1 = 0.012 \text{ m}^3, \quad V_s =$ volume at s.t.p.

$T_1 = 288 \text{ K}, \quad T_2 = 273 \text{ K}$

gives
$$V_s = \frac{P_1}{P_s} \times \frac{T_s}{T_1} \times V_1$$

$$= \frac{763.13}{760} \times \frac{273}{288} \times 0.0126 \text{ m}^3$$

$$= 0.0120 \text{ m}^3$$

Now,

Energy released by gas = energy gained by water

$$= 3.7 \text{ [kg]} \times (294 - 285) \text{ [K]} \times 4.18 \left[\frac{\text{kJ}}{\text{kg K}}\right]$$

$$= 139.2 \text{ kJ}$$

236 Mechanical engineering science

But,

Energy released by gas = volume × higher calorific value

$$\therefore \text{H.C.V.} = \frac{139.2}{0.012} \frac{[\text{kJ}]}{[\text{m}^3]}$$

$$= 11\ 600\ \text{kJ/m}^3$$

Mass of condensate collected $= 1.6\ \text{g} = 1.6 \times 10^{-3}\ \text{kg}$

$$\therefore m_c = \frac{1.6 \times 10^{-3}}{0.012} \frac{[\text{kg}]}{[\text{m}^3]}$$

$$= 0.133\ \text{kg/m}^3$$

Taking h_{fg} as 2442 kJ/kg then in equation (12.1),

$$\text{L.C.V.} = \text{H.C.V.} - m_c \times h_{fg}$$

$$= 11\ 600\ \left[\frac{\text{kJ}}{\text{m}^3}\right] - 0.133\ \left[\frac{\text{kg}}{\text{m}^3}\right] \times 2442\ \left[\frac{\text{kJ}}{\text{kg}}\right]$$

$$= 11\ 275\ \text{kJ/m}^3$$

The higher calorific value of the gas is 11 600 kJ/m³ and its lower calorific value is 11 275 kJ/m³.

Problems

1. A sample of coal has the following analysis by mass:

Carbon 88%,   Hydrogen 9%,   Impurities 3%

Calculate the mass of air required for the complete combustion of 0.22 kg of the coal.

2. Calculate the minimum quantity of air required for the complete combustion of 1 kg of benzene, C_6H_6.

3. The percentage analysis by mass of a fuel is:

Carbon 76%,   Hydrogen 11%,   Oxygen 7%,   Incombustibles 6%

Calculate the minimum quantity of air required for the complete combustion of 1 kg of the fuel and the percentage analysis by mass of the total combustion products.

4. A fuel oil has the following percentage analysis by mass:

Carbon 83.6%,   Hydrogen 12.8%,   Oxygen 1.4%,   Nitrogen 2.2%

Combustion 237

Determine the minimum quantity of air required for the complete combustion of 0.63 kg of the oil.

5. Calculate the theoretical air to fuel ratio by mass for the complete combustion of octane, C_8H_{18}, and the percentage analysis by mass of the dry combustion products.

6. The coal supplied to a boiler has the following composition by mass: Hydrogen 4%, Carbon 84%, Moisture 5%, remainder being ash
If the air supplied is 40% in excess of that required determine the analysis of the total combustion products by mass and by volume.

7. A sample of coal has the following composition by mass:
Carbon 81%, Hydrogen 8%, Oxygen 6%, Incombustibles 5%
Find the minimum mass of air required for the complete combustion of 1 kg of the fuel.
If 30% excess air is supplied, find the percentage composition of the dry flue gas by mass and by volume.

8. The analysis by mass of the petrol supplied to an engine is 84% carbon and 16% hydrogen. Calculate the chemically correct air/fuel ratio.
If the actual air/fuel ratio supplied is 17.5:1 determine the mixture strength and the excess air supplied per kg of fuel.

9. The volumetric analysis of a sample of coal is:
Hydrogen 48%, Carbon monoxide 16%, Methane 30%, Nitrogen 6%
Determine:
(a) the volume of air required for the combustion of 200 m³ of the gas;
(b) the analysis by mass of the total combustion products.

10. A fuel contains 84% carbon, 12% hydrogen, and 2% oxygen by mass, with the remainder being incombustible. Calculate the minimum quantity of air required for complete combustion, and the mass of steam and carbon dioxide produced per kg of fuel.

11. A sample of producer gas has the following analysis by volume:

Hydrogen 15%, Carbon monoxide 20%, Carbon dioxide 4%
Methane 11%, Oxygen 7%, Nitrogen 43%

Calculate the volume of air required for the complete combustion of 250 l of the gas and the volumetric analysis of the dry combustion products.

238 Mechanical engineering science

12. A sample of petrol has the following percentage analysis by mass: carbon 86% and hydrogen 14%. Calculate the minimum mass of air required for the complete combustion of 0.5 kg of the petrol.

If during the combustion the actual air to fuel ratio is 10:1 calculate the mixture strength as a percentage rich or weak mixture.

13. Distinguish between higher and lower calorific value as applied to a fuel.

The analysis by mass of a coal is:

Carbon 84%, Hydrogen 3%, Oxygen 2%, Incombustibles 11%

Calculate the higher and lower calorific values of the fuel.

Assume the calorific values to be 34 000 kJ/kg for carbon and 161 000 kJ/kg for hydrogen and take the condensate allowance as 2335 J/g.

14. A gas fired boiler is supplied with a gas having an analysis by volume:

    Carbon monoxide 12%, Methane 19%, Hydrogen 22%
    Carbon dioxide 9%, Nitrogen 38%

Calculate the Higher Calorific Value of the fuel in kJ/m^3 and the minimum volume of air required for the complete combustion of 3 m^3 of the gas.

The calorific values of the combustible constituents are:

    Carbon monoxide, 11 850 kJ/m^3; Methane, 37 750 kJ/m^3;
    Hydrogen, 11 900 kJ/m^3.

15. The coal supplied to a boiler has the following percentage analysis by mass:

    Carbon 86%, Hydrogen 5%, Oxygen 3%, Sulphur 1%,
    Incombustibles 5%

Calculate the higher and lower calorific values of the fuel.

    (Calorific value of carbon   =   33 600 kJ/kg
    Calorific value of sulphur   =   9 060 kJ/kg
    Calorific value of hydrogen = 161 000 kJ/kg
    Condensate allowance   =   2 440 kJ/kg)

16. In a test carried out on a bomb calorimeter to determine the calorific value of a coal the following results were obtained.

    Mass of coal used   = 1.05 g
    Water equivalent of apparatus = 350 g
    Mass of Water   = 1.400 kg
    Temperature rise (corrected) = 4.65°C

Calculate the calorific value of the coal.

17. An experiment to determine the calorific value of a water gas, using a gas calorimeter, yielded the following results:

Barometric pressure	= 754 mm mercury
Gas Pressure	= 107 mm water above atmospheric
Gas Temperature	= 10°C
Volume of gas consumed	= 0.0084 m³
Cooling water collected	= 2.58 kg
Cooling water inlet temperature	= 8.9°C
Cooling water outlet temperature	= 13.7°C

Determine the calorific value of the gas in kJ/m^3, measured at s.t.p. conditions. (Relative density of mercury is 13.6.)

13

Heat transfer

13.1 Modes of heat transfer

The transfer of thermal energy from a hot body to one at a lower temperature can arise from conduction, convection, or radiation. These modes of transfer can occur singly within a system or simultaneously.

13.2 Conduction

Conduction is the transfer of thermal energy between different parts of the same substance or between two materials in physical contact. There is little movement of the atoms or molecules of a substance and energy is passed on from one atom or molecule to the next, etc. Thus, metals, with a close molecular structure, are better thermal conductors than liquids and gases, which have an increasingly greater molecular dispersal and freedom respectively.

13.3 Convection

Convection is the transfer of thermal energy resulting from the movement of a fluid. Hence the process is only applicable to liquids and gases. Convection can occur naturally as a result of a temperature variation, and hence a density variation, or it can be forced by mechanical means.

An example of natural convection is the air currents created in a room by the presence of a radiator. The air in contact with the radiator experiences a temperature increase, expands and so becomes less dense and rises. This causes an upward air movement at the radiator and a general movement of air throughout the room results. An example of

Heat transfer 241

forced convection is the use of a hair dryer where an electrically operated fan is used to force air over a hot coil.

Whenever heat transfer from a solid to a liquid or to a gas occurs it takes place partly by conduction and partly by convection. Consider the case of an immersion heater. The electrically heated coils are inside a metal tube and therefore the energy is initially transferred to the water in contact with the metal tubes by conduction. This water then expands, becomes less dense, and creates a convection current to disperse the heat throughout the water.

13.4 Radiation

Radiation is the transfer of thermal energy by means of electro-magnetic waves. These are similar to light waves, the difference being in the frequency. Unlike conduction and convection, no medium is required for the transfer of radiant energy and the best example of this form of heat transfer is the energy which the earth receives from the sun—through a vacuum. Again contrasting with conduction and convection, where the temperature difference mainly determines the rate of heat transfer, radiation is dependent on the temperature level.

Thermal radiation travels in straight lines, unless impeded by an intervening body, and its waves are reflected from a polished surface in the same way as for light waves. The thermal radiation emitted or absorbed by a body also depends upon whether the surface is light or dark. Dark surfaces emit, and absorb, much more radiant energy than light surfaces.

13.5 Coefficient of thermal conductivity

The heat energy transferred by conduction has been found to depend upon:

1. the material;
2. the thickness of the material;
3. the surface area through which conduction can take place;
4. the temperature difference across the thickness of the material;
5. the time taken.

The heat transfer will be directly proportional to the area, the temperature difference and the time taken. However, if the thickness is increased the heat transfer will decrease. Hence the heat transfer is inversely proportional to the thickness of material. The value of the coefficient of thermal conductivity of a material, k, takes these factors into account and is defined as the heat energy transferred per unit area, per unit thickness, per unit time, per unit temperature difference. The units of k are therefore $J\,m/m^2\,s\,K$ or $W/m\,K$.

A material having a high thermal conductivity is a good conductor of heat, whereas one having a low thermal conductivity is a poor conductor of heat. Generally materials which are good thermal conductors are also good electrical conductors, and vice versa.

Average values of the coefficient of thermal conductivity are given in the following table.

Table 5. Average values of the coefficient of thermal conductivity at 300 K

MATERIAL	COEFFICIENT OF THERMAL CONDUCTIVITY (k, W/m K)
Copper	386
Aluminium	202
Duralumin	160
Brass	120
Cast iron	49
Mild Steel	55
Lead	36
Glass	0.8 –1.1
Concrete	0.9 –1.4
Brick	0.35–0.7
Asbestos	0.163
Wood	0.02
Felt	0.041
Cork	0.044

13.6 Conduction through a material

The rate of heat energy transfer through a material is given by

$$Q = \frac{kAT}{x} \text{ watts} \tag{13.1}$$

where A = conducting area [m²]

   T = temperature difference between conducting faces [K or °C]

   x = thickness of material [m]

and k = coefficient of thermal conductivity of the material [W/m K or W/m °C]

Example 13.1

A glass window measures 1.3 m × 1 m and the thickness of the glass is 4 mm. If the temperature of the inside surface of the glass is 10°C, and that of the outside surface is 5.6°C calculate the rate of heat transfer through the window. (For glass, coefficient of thermal conductivity = 0.95 W/m K.)

Solution

From equation (13.1) the rate of heat transfer through the window is given by

$$Q = \frac{kAT}{x}$$

where $k = 0.95$ W/m K
$A = 1.3 \times 1 = 1.3$ m²
$T = (10 - 5.6)°C = 4.4$ K
$x = 4$ mm $= 0.004$ m

Then
$$Q = \frac{0.95 \text{ [W/m K]} \times 1.3 \text{ [m}^2\text{]} \times 4.4 \text{ [K]}}{0.004 \text{ [m]}}$$

$= 1358$ W

The rate of heat transfer through the window is 1358 W.

13.7 Conduction through a plane composite wall

Consider a composite wall as shown in Fig. 13.1 where the temperatures at the surfaces are t_1, t_2, t_3, and t_4 respectively and the material

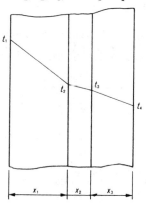

13.1 Conduction through a plane composite wall

thicknesses are x_1, x_2, and x_3 respectively, as shown. Assuming that the cross-sectional area of the wall is uniform and that the coefficients of thermal conductivity for the three materials are k_1, k_2, and k_3 respectively, then:

Rate of energy transfer through material 1 is

$$Q_1 = \frac{k_1 A(t_1 - t_2)}{x_1}$$

Similarly

$$Q_2 = \frac{k_2 A(t_2 - t_3)}{x_2}$$

and

$$Q_3 = \frac{k_3 A(t_3 - t_4)}{x_3}$$

However for equilibrium, $Q_1 = Q_2 = Q_3 = Q$ as if this condition does not exist the temperatures t_1, t_2, t_3, and t_4 could not remain stable.
Rewriting the above equations gives

$$t_1 - t_2 = \frac{Q x_1}{k_1 A}$$

$$t_2 - t_3 = \frac{Q x_2}{k_2 A}$$

$$t_3 - t_4 = \frac{Q x_3}{k_3 A}$$

Adding these equations gives

$$t_1 - t_4 = \frac{Q}{A}\left(\frac{x_1}{k_1} + \frac{x_2}{k_2} + \frac{x_3}{k_3}\right)$$

or

$$Q = \frac{A(t_1 - t_4)}{\left(\frac{x_1}{k_1} + \frac{x_2}{k_2} + \frac{x_3}{k_3}\right)} \tag{13.2}$$

Example 13.2

A brick wall of thickness 225 mm is faced with concrete, 25 mm thick, on the outside and plaster, 12 mm thick on the inside. Calculate the heat flow rate through the wall if it is 6 m long and 2.5 m high when

Heat transfer

the temperature of the inside surface is 21°C and that of the outside is 7.2°C. What are the interface temperatures?

(Thermal conductivity of brick = 0.52 W/m K
Thermal conductivity of concrete = 0.865 W/m K
Thermal conductivity of plaster = 0.45 W/m K)

Solution
A diagram is given in Fig. 13.2.

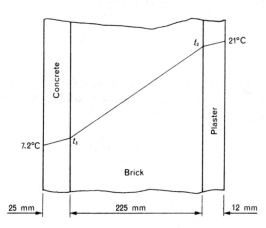

13.2 Example 13.2

Let the temperatures at the interfaces be t_2 and t_3 as shown. Applying equation (13.2)

$$Q = \frac{A(t_1 - t_4)}{\left(\frac{x_1}{k_1} + \frac{x_2}{k_2} + \frac{x_3}{k_3}\right)}$$

where $A = 6 \times 2.5 = 15 \text{ m}^2$

$t_1 = 21°C$

$t_4 = 7.2°C$

$x_1 = 12 \text{ mm} = 0.012 \text{ m}$ $x_2 = 0.225 \text{ m}$, $x_3 = 0.025 \text{ m}$;

$k_1 = 0.45 \text{ W/m K}$, $k_2 = 0.52 \text{ W/m K}$, $k_3 = 0.865 \text{ W/m K}$.

Then,
$$Q = \frac{15(21-7.2)}{\left(\frac{0.012}{0.45} + \frac{0.225}{0.52} + \frac{0.025}{0.865}\right)}$$

$$= \frac{15 \times 13.8}{(0.0267 + 0.433 + 0.0289)}$$

$$= \frac{207}{0.4886} = 424 \text{ W}$$

To determine the interface temperatures we shall now consider the rate of heat transfer through the individual materials. Thus, for plaster, since the heat transfer rate must be equal to the overall heat transfer rate, then:

$$Q = \frac{Ak_1(t_1 - t_2)}{x_1}$$

$$424 = \frac{15 \times 0.45(21 - t_2)}{0.012}$$

$$(21 - t_2) = \frac{424 \times 0.012}{15 \times 0.45}$$

$$= 0.75$$

$$\therefore t_2 = 20.25°\text{C}$$

Similarly for the brick:

$$Q = \frac{Ak_2(t_2 - t_3)}{x_2}$$

$$424 = \frac{15 \times 0.52(20.25 - t_3)}{0.225}$$

$$20.25 - t_3 = \frac{424 \times 0.225}{15 \times 0.52} = 12.23$$

$$\therefore t_3 = 8.02°\text{C}$$

Alternatively, the interface temperature t_3 could have been obtained by considering the heat transfer rate through the concrete, i.e.

$$Q = \frac{Ak_3(t_3 - t_4)}{x_3}$$

$$424 = \frac{15 \times 0.865(t_3 - 7.2)}{0.025}$$

$$t_3 - 7.2 = \frac{424 \times 0.025}{15 \times 0.865} = 0.82$$

$$t_3 = 8.02°C$$

The rate of heat transfer through the wall is 424 W and the interface temperatures are 20.25°C and 8.02°C respectively.

13.8 Stefan–Boltzman law for radiation

A body which absorbs all the radiant energy falling upon it is known as a black body. This in no way implies that its colour is black. For example, snow is almost perfectly black with respect to thermal radiation but owes its colour to the fact that it reflects almost all the light falling upon it. In practice no totally black body exists.

The energy emitted from a body per unit area per unit time is known as the radiant emittance, M, of that body and the Stefan–Boltzman law for radiation states that the radiant emittance of a black body is proportional to the fourth power of its thermodynamic temperature. Thus,

$$M = \sigma T^4 \qquad (13.3)$$

where σ = constant = 56.7×10^{-12} kW/m² K⁴.

Consider a black body at temperature T_1 to be surrounded by black surroundings which are at a lower temperature T_2.

The radiant emittance from the black body = σT_1^4

This energy will be completely absorbed by the black surroundings. Similarly,

The radiant emittance from the black surroundings = σT_2^4

Now, as the body is black it will absorb the energy emitted by the black surroundings. The rate of energy transfer per unit area or radiance, L, from the black body to the surroundings is given by

$$L = \sigma(T_1^4 - T_2^4) \qquad (13.4)$$

Example 13.3

The rate at which energy from the sun normally falls upon the surface of the earth is 1.4 kW/m². If the radius of the sun is 7.25×10^5 km and its distance from the earth is 14.8×10^7 km estimate the surface temperature of the sun.

Stefan–Boltzman constant = 56.7×10^{-12} kW/m² K⁴

Solution

Assume the thermodynamic temperature of the surface of the sun to be T. Then,

$$\text{Radiant emittance of sun} = \sigma T^4 \quad \text{(equation 13.3)}$$

$$= \frac{56.7}{10^{12}} \times T^4 \text{ kW/m}^2$$

$$\text{Radius of sun} = 7.25 \times 10^5 \text{ km}$$

$$= 7.25 \times 10^8 \text{ m}$$

$$\therefore \text{ Surface area of the sun} = 4\pi \times (7.25 \times 10^8)^2 \text{ m}^2$$

$$\therefore \text{ Radiant intensity of sun} = \frac{56.7}{10^{12}} \times T^4 \times 4\pi \times (7.25 \times 10^8)^2 \text{ kW}$$

Now the earth can be considered as a small surface on an imaginary sphere of radius 14.8×10^7 km (14.8×10^{10} m), the sun being at the centre of this sphere.

$$\text{Surface area of imaginary sphere} = 4\pi \times (14.8 \times 10^{10})^2 \text{ m}^2$$

$\therefore$ Irradiance on surface of earth

$$= \frac{56.7 \times T^4 \times 4\pi \times (7.25 \times 10^8)^2}{10^{12} \times 4\pi \times (14.8 \times 10^{10})^2} \text{ kW/m}^2$$

But the irradiance on the earth's surface is 1.4 kW/m².

$$1.4 = \frac{56.7 \times T^4 \times 4\pi \times (7.25 \times 10^8)^2}{10^{12} \times 4\pi \times (14.8 \times 10^{10})^2}$$

$$T^4 = \frac{1.4 \times 14.8^2 \times 10^{16}}{56.7 \times 7.25^2}$$

$$T = 5663 \text{ K} = 5390°\text{C}$$

The surface temperature of the sun is 5663 K or 5390°C.

Example 13.4

Two copper spheres have diameters in the ratio 1:2. The smaller sphere cools down at a rate of 4°C/min when its temperature is 177°C and the temperature of the surrounding atmosphere is 27°C. At what rate would the larger sphere cool down in the same surroundings if its own temperature is 157°C?

Heat transfer

Compare the initial rates of energy transfer from the spheres to the surrounding atmosphere.
Assume the Stefan–Boltzman law to be applicable.

Solution

The rate of energy transfer per unit area, or radiance, L, from the spheres to the surrounding atmosphere is given by

$$L = \sigma(T_1^4 - T_2^4) \qquad \text{(equation 13.4)}$$

where T_1 = temperature of sphere

T_2 = temperature of surrounding atmosphere

and σ = Stefan–Boltzman constant

Denoting the large and small spheres by suffices l and s respectively—
For small sphere:

$$T_{1s} = 177 + 273 = 450 \text{ K}; \qquad T_{2s} = 27 + 273 = 300 \text{ K}$$
$$\therefore L_s = \sigma(450^4 - 300^4) \text{ kW/m}^2$$

Assuming the radius of this sphere to be r then its area is $4\pi r^2$.

$\therefore$ Rate of energy transfer, $P_s = \sigma(450^4 - 300^4) \times 4\pi r^2$ kW

For large sphere:

$$T_{1l} = 157 + 273 = 430 \text{ K}; \qquad T_{2l} = T_{2s} = 300 \text{ K}$$
$$\therefore L_l = (430^4 - 300^4) \text{ kW/m}^2$$

The radius of this sphere will be $2r$, i.e. a surface area of $4\pi \times (2r)^2$.

$\therefore$ Rate of energy transfer, $P_l = \sigma(430^4 - 300^4) \times 4\pi \times (2r)^2$ kW

$$\therefore \frac{P_l}{P_s} = \frac{\sigma(430^4 - 300^4) \times 4\pi \times (2r)^2}{\sigma(450^4 - 300^4) \times 4\pi r^2}$$

$$= \frac{(430^4 - 300^4)}{(450^4 - 300^4)} \times 4$$

$$= 3.17$$

The initial rate of energy transfer from the larger sphere is 3.17 times that from the smaller sphere. But,

Rate of energy transfer from sphere = mass × specific heat capacity
× temp. fall rate
$$= mC\theta$$

If the density of the material is ρ and its specific heat capacity is C, then

$$m_s = \tfrac{4}{3}\pi r^3 \rho; \qquad m_1 = \tfrac{4}{3}\pi(2r)^3 \rho$$

$\theta_s = 4°C/\text{min}; \qquad \theta_1 = $ required initial temp. fall rate of large sphere

$$\therefore \frac{P_1}{P_s} = \frac{\tfrac{4}{3}\pi(2r)^3 \rho C \theta_1}{\tfrac{4}{3}\pi r^3 \rho C \times 4} = 3.17$$

$$2\theta_1 = 3.17$$

$$\theta_1 = 1.585°C/\text{min}$$

The initial rate of fall in temperature of the large sphere is 1.585°C/min.

Problems

1. A brick wall is 0.24 m thick and the temperature difference between its surfaces is 14 K. Calculate the heat flow rate through the wall if it has a length of 8 m and a height of 2.15 m. The thermal conductivity of brick is 0.61 W/m K.

2. A plate glass window is 6 mm thick. If the loss by conduction is 810 W/m² calculate the temperature difference between the surfaces of the glass. Assume the coefficient of thermal conductivity of the glass to be 0.9 W/m K.

3. A concrete wall has a thickness of 0.25 m and a surface area of 8 m². Calculate the rate of heat transfer through the wall when the temperature difference between the inside and outside surfaces is 8°C. Assume the thermal conductivity to be 0.9 W/m K.

4. (*a*) Describe a method of determining the thermal conductivity of a good conductor of heat. Give a sketch of the apparatus.
(*b*) The outside of a boiler of surface area 30 m² has a lagging 40 mm thick. The outer surface of the lagging has a temperature of 50°C, and the mean temperature of the boiler plate is 200°C. Calculate the heat flow rate through the lagging. (k for the lagging $= 0.74$ W/m K.)

5. A duralumin tank has a wall thickness of 10 mm. Calculate the heat flow rate through the tank, when the inside and outside surfaces of the tank are at temperatures of 80°C and 18°C respectively.

If the outside wall of the tank is now covered with an asbestos sheet of thickness 5 mm so that the outside surface of the asbestos is at a temperature of 18°C what would then be the rate of heat transfer through the material?
For duralumin $k=210$ W/m K; For asbestos $k=0.19$ W/m K.

6. A wall is 5 m long and 2.8 m high and the temperature of the inside and outside surfaces are 22°C and 15°C respectively. Calculate the heat flow rate through the wall when:

(a) it is constructed of two layers of bricks each 115 mm thick, so that the total thickness is 230 mm;

(b) it is constructed as in (a) but with a 50 mm air gap between the two layers of bricks.

Assume the coefficient of thermal conductivity to be 0.725 W/m K for the brick and 0.026 W/m K for air.

7. Determine the heat flow rate through a steel plate 7 mm thick when the temperatures of the plate surfaces are 104°C and 88°C respectively.

If the plate is covered by insulating material so that the heat flow rate is reduced by 54%, what would then be the temperature difference between the plate surfaces? (Assume k for steel $=46$ W/m K.)

8. A tiled roof has an effective thickness of 50 mm and the inside of the roof is lined with felt of thickness 3 mm. When the temperature difference between the inside and outside surfaces is 8 K what will be the heat flow rate for a roof area of 180 m²?

What would be the percentage increase in the heat conducted if the felt lining were omitted?
For the tiles, $k=0.52$ W/m K; for felt, $k=0.052$ W/m K.

9. Explain what is meant by the thermal conductivity of a material.

Steam is passed through a glass tube which is surrounded by a water-jacket. The external diameter of the tube is 10 mm, its wall thickness is 0.8 mm and the length of the water-jacket is 0.3 m. The inlet temperature of the water is 14°C and its outlet temperature is 20°C when the flow rate is 1.80 litres per minute. Determine the mean value for the thermal conductivity of glass.

10. The window in a room has an area of 3 m² and a thickness of 2.5 mm. Assuming that the only heat losses from the room occur by conduction through the window find the cost per day of maintaining a room temperature of 20°C when the outside air temperature is 10°C.

252 Mechanical engineering science

What saving would be achieved by double glazing the window with a similar sheet of glass and separated by 10 mm of air from the first?

Thermal conductivity of glass = 0.93 W/m K
Thermal conductivity of air = 0.026 W/m K
Cost of energy   = 0.73 p/kWh

11. A compound wall consists of a layer of firebrick 36 mm thick and a layer of asbestos 22 mm thick. Determine the temperature at the interface when the temperature at the outside of the firebrick is 450°C and that of the outer surface of the asbestos is 15°C. What thickness of asbestos is required to ensure that the temperature drop across the firebrick does not exceed 60°C?

Thermal conductivity of asbestos = 0.124 W/m K
Thermal conductivity of firebrick = 0.84 W/m K

12. A solid copper sphere cools down at a rate of 5°C/min when its temperature is 227°C and the temperature of the surrounding atmosphere is 27°C. At what rate would a similar solid copper sphere of four times the diameter cool down in the same surroundings when its temperature is 277°C? Assume conditions are such that the Stefan–Boltzman law can be applied.

13. Assuming that the earth acts as a perfect black body and that the temperature of the surrounding space is 0 K calculate the average temperature of the surface of the earth. The intensity of radiation from the sun on the earth's surface is 1.4 kW/m²,

Stefan–Boltzman constant = 56.7×10^{-12} kW/m² K⁴

14. A solid iron ball has a diameter of 40 mm. It is heated in a furnace until its temperature is 607°C and then placed in an atmosphere whose temperature is 17°C. Assuming the ball to act as a perfect black body determine the initial rate of cooling of the ball.

Stefan–Boltzman constant   = 56.7×10^{-12} kW/m² K⁴
Density of iron   = 7.9 Mg/m³
Specific heat capacity of iron = 0.476 kJ/kg K

Answers to Problems

Chapter 2 (page 27)

1. 2458 N; 4121 N
2. 142.3 kN; 289.4 kN
3. 185.6 N at 78° below force of 260 N
4. 139 N at 319°
5. 120 N at 167°, 80 N at 292° or 120 N at 250°, 80 N at 125°. Angles measured in same direction as 50° angle.
6. (a) 26.2 N, (b) 94.4 N at 13° 54′ to AD
7. 687 N at wall; 2079 N at 70° 42′ to the ground
8. 42.5 N; 127.5 N; 374 N
9. 512.2 N; 512.2 N at 50° to AB
10. $AD = BE = -4386$ N; $DC = CE = 4121$ N; $DE = 0$
11. Forces in kN: $AB = BC = 12$; $CD = -20.8$; $BD = 0$; $AD = 2.5$; $DE = -25.1$
12. Forces in kN: $AE = BF = 17.3$; $CF = -20$; $EF = -30$; $DE = 46.2$; $CD = -40.4$
13. Forces in kN: $AB = 17$; $BC = 6$; $CD = DE = -5.2$; $BD = 4$; $BE = -11$
14. (1) = 50 kN; (2) = -26.3 kN; (3) = (4) = -50.8 kN
15. Forces in kN; $AB = 70.5$; $BC = 48$; $CD = -41.6$; $BD = -39$; $AD = -31.2$; $DE = -90.9$
16. Forces in kN: $R_A = 13$; $R_F = 11$; $AB = BC = CD = -13$; $DE = EF = -11$; $GH = 11$; $BH = 18.4$; $DH = 2.83$; $DG = -2$; $GE = 15.5$; $AH = CH = FG = 0$
17. Forces in kN: $R_A = 60$; $R_D = 50$; $AB = -69.3$; $BC = -52$; $CD = -57.7$; $DE = EF = 28.85$; $FG = GA = BF = 34.65$; $BG = 30$; $CF = 46.2$; $CE = 0$
18. $R_1 = 38\,000$ N; $R_2 = 32\,000$ N; (1) = $-53\,740$ N; (2) = 23 324 N; (3) = 11 662 N; (4) = $-45\,254$ N; (5) = 32 000 N; (6) = 26 000 N; (7) = 38 000 N

254 Mechanical engineering science

19. Forces in kN: $R_A = 17.16$; $H_A = 13.42$; $R_E = 24.66$; AB $= -38.35$; BC $= -29.35$; CD $= -18.38$; DE $= -29.6$; EF $= 16.44$; FG $= 34.29$; GH $=$ HA $= 47.7$; HB $= 0$; BG $= -15$; GC $= 6.64$; CF $= -25.24$; FD $= 17.85$.
20. Forces in kN: $R_A = 15.7$; $R_C = 16.3$; AB $= -38.63$; CB $= -40.06$; EB $= 17.23$; DB $= 19.92$; AE $= 31.05$; ED $= 21.1$; DC $= 32.28$.

Chapter 3 (page 55)

1. 66.2 kg; 730 N
2. (a) 99.4 kg, (b) 1000 N
3. (a) 2263 N, 1505 N, (b) 10.65 m from P
4. (a) $R_A = 3174$ N, $R_C = 19\ 550$ N; (b) 13 740 N
5. 1510 N; 923 N at 11° to the beam
6. $H_A = 101.4$ N; $V_A = 15.5$ N; $H_C = 198.6$ N; $V_C = 164.5$ N
7. 300 kg; 5.24 kN at 13° 50′ to the beam
8. $R_A = 220$ N; $R_B = 149$ N; $R_C = 273$ N
9. $R_A = 166$ N at 62° to AE; $R_E = 105$ N at 41° 30′ to AE
10. 26.6 N m
11. 225 N
12. 76.8 N m
13. 5.07 mm to right of YY; 1.52 mm below XX.
14. 63.17 mm from side of 0.10 m; 36.58 mm from side of 0.18 m
15. (a) 100 mm, (b) 118 mm
16. $\bar{x} = 30.4$ mm; $\bar{y} = 58.44$ mm; $\bar{z} = 98.6$ mm
17. $\bar{x} = 68.2$ mm; $\bar{y} = 72.5$ mm; $\bar{z} = 189.8$ mm
18. 6.16 mm (5.2 mm parallel to side of 350 mm and 3.3 mm parallel to side of 240 mm); 109.6 mm from the centre parallel to the side of 350 mm and 69.4 mm from the centre parallel to the side of 240 mm

Chapter 4 (page 74)

1. (a) 95 833 kN/m² (b) 0.000 483 (c) 198.4 GN/m² (d) 3.34
2. 20.7 mm; 0.857 mm
3. 0.01 m
4. Yes, they have equal values for E
5. AB, $\sigma = 95.5$ MN/m²; BC, $\sigma = 48$ MN/m²; 0.162 mm
6. Length of 30 mm square section is 0.523 m
   For 30 mm section, $\sigma = 47$ MN/m²; For 17.5 mm section, $\sigma = 139$ MN/m²
7. (a) 212.5 GN/m² (b) 331 MN/m² (c) 555 MN/m² (d) 31% (e) 42%
8. 1.335 mm

Answers to problems 255

9. 19.2 mm
10. 65.5 kN; 39.3 MN/m^2; 11.95
11. 202 GN/m^2; No
12. (a) 210 GN/m^2 (b) 465 MN/m^2 (c) 33% (d) 32.7%
13. (a) 196 MN/m^2 (b) 105 GN/m^2 (c) 360 MN/m^2 (d) 63% (e) 67.6%
14. 5.96 kN
15. 17.6 MN/m^2; 15.7 MN/m^2
16. 106 kN
17. 67.6 MN/m^2; 17.7
18. 148°

Chapter 5 (page 90)

1. 1189 N
2. 0.33; (a) 8312 N, (b) 15 768 N
3. (a) 0.276, (b) 2578 N
4. 1108 N
5. 931 N
7. 74.8 N
8. 0.312
9. 20 kg
10. (a) 36.7 N m, (b) 115.3 kJ
11. 0.034
12. 13.77 N m; 64.9 kJ

Chapter 6 (page 109)

1. 208 N
2. 19.08 kN
3. 271 N; 323 J
4. (a) 52.36, (c) $E = 0.05W + 2.9$, (d) 38.2%
5. 19.6%
6. (a) 39.9, (b) 1555 J, (c) 52.1 N
7. (a) 29.5%, (b) 35%
8. (a) 29.6%, (b) 37.3%
9. $E = 0.045W + 8.8$; 32.6%
10. 71.5%
11. $E = 0.0865W + 42$; (a) 1296 N, (b) 32.1%
12. 3385 N; 1777 J

13. 40%
14. 465 N
15. 179 N
16. $E = 0.08W + 32.8$; 432.8 N; 81%; 660 N

Chapter 7 (page 131)

1. 1.11 m/s^2; 222.2 m
2. (a) 0.375 m/s^2, (b) 6 min 40 s
3. 49.8 s
4. (a) 101 km/h, (b) 1.473 km
5. (a) 34.4 m, (b) 92 m
6. 220 m; 2.17 m/s^2
7. (a) 8 km, (b) 0.104 m/s^2, (c) 0.185 m/s^2
8. 248 m
9. (a) 0.33 m/s^2, (b) 3 min 9 s
10. 1.32 km; 6.27 km
11. (a) 39.6 m/s, (b) 4.04 s
12. (a) 20.51 km, (b) 55.1 s, (c) 3727 m
13. 20° 9′ to the horizontal; 2.25 s
14. (a) 28.53 km, (b) 159 s, (c) 88.7 km
15. (a) 7.95 km/h in a direction 28° N of W, (b) 4.41 km
16. 637 km/h in a direction 41° 40′ N of W; 2.03 km
17. 136 km/h in a direction 34° 12′ W of N
18. (a) 39° W of N, (b) 15.03 km; 6 min 4s
19. (a) 9.6 km/h due south, (b) 4.8 km
20. 22.3 km/h in a direction 14° 15′ N of E; 7.76 km

Chapter 8 (page 148)

1. 3556 N
2. (a) 4917 N, (b) 26 666 kg m/s
3. (a) 0.168 m/s^2, (b) 50.64 kN, (c) 8.33×10^6 kg m/s
4. 0.682 m/s^2
5. 298.3 N
6. (a) 0.642 m/s^2, (b) 12.48 s, (c) 2641 J, 7923 J
7. (a) 33.43 kN, (b) 1 min 51 s
8. (a) 183 N, (b) 195 N
9. 0.327 m/s^2; 47 s
10. (a) 0.0613 m/s^2, (b) 103.3 km/h
11. 28.64 m
12. (a) 163.5 N parallel to plane, (b) 0.0327 m/s^2, (c) 36.78 kJ

Answers to problems 257

13. (a) 24.57 kN, (b) 2.95
14. 10.55 kN
15. (a) 62.4 kN, (b) 49.8 km/h, (c) 4 min 35 s
16. 97 mm
17. (a) 2.26 m/s^2, (b) 60.4 N
18. (a) 0.613 m/s^2, (b) 31.3 N, (c) 2.55 s
19. (a) 8.24 N, (b) 0.654 m/s^2, (c) 1.308 m/s
20. 0.1 kg; 39.7 N
21. 1.065 kg; 63.66 N
22. (a) 7.65 km/h, (b) 34.2 m
23. 8.86 km/h
24. 0.286 m/s in direction of second truck
25. (a) 15.85 m/s^2, (b) 1.342 MN
26. 18 mm
27. 3.17 m
28. 2.76 kN
29. 30 m/s

Chapter 9 (page 171)

1. (a) 0.168 rad/s^2, (b) 88.9 m
2. (a) 209.4 rad/s, (b) 31.4 m/s, (c) 2.094 rad/s^2, (d) 0.314 m/s^2
3. 16.98 rad/s^2; 4.71 s
4. (a) 5.05 rad/s^2, (b) 25.7 rev, (c) 44.44 m
5. (a) 287.7 rev/min, (b) 61.5 rev, (c) 0.544 m/s^2
6. (a) 7.07 m/s, (b) 2.618 rad/s^2, (c) 188.5 rad
7. (a) 1.4 rad/s^2, (b) 79.2 m
8. (a) 0.419 rad/s^2, (b) 82.5 rev
9. 137 m/s^2
10. 128.6 m/s^2; 28.28 km/h
11. 31.73 rev/min
12. (a) 85.9 N, (b) 192.3 rev/min
13. 1.24 kg at 224° to mass of 0.8 kg, measured in the same direction as the 90°
14. (a) 2857 N, (b) at a radius of 0.543 m at 289° to mass of 5 kg
15. At a radius of 0.323 m at 275° to the mass of 3 kg, measured in the same direction as the 130°
16. (a) 13.4 m/s, (b) 12 m/s, (c) 55.3 rad/s
17. (a) 45.75 rad/s; (b) 38.88 m/s at top, 0 at bottom, 27.5 m/s at ends of horizontal diameter
18. (a) 8.23 m/s; 10.7 m/s; 13.1 m/s
19. 1.46 m/s; 0.855 rad/s; 6.63 rad/s

Chapter 11 (page 209)

1. 120.6°C
2. 8.062 kN
3. 190.6°C
4. 17.77 m^3
5. 98.039 kN/m^2; 96.11 kN/m^2
6. 0.0388 m^3
7. (a) 49°C, (b) 614.2 kN/m^2, (c) 0.08 m^3
8. 142.2 J/kg K; 0.706 kg
9. 1.516 kg; 137.5°C
10. 256 J/kg K; 926 J/kg K
11. (b) 820 kN/m^2; 729 kN/m^2
12. 0.239 m^3; (a) 0.224 m^3, (b) 478 kN/m^2
13. 3.22 kg
14. 39 kg
15. (a) 32, (b) 893.5 J/kg K
16. 55.53 kN/m^2
17. 5.77 kN/m^2; 7.1 kg
18. 1410 kN/m^2
19. 97.37 kN/m^2; 202.63 kN/m^2; 222.8 J/kg K
20. 2.629 MN/m^2
21. 0.664 kg O_2; 2.187 kg N_2; 75.72 kN/m^2; 284.28 kN/m^2; 0.022 kg; 40 kN/m^2
22. 87.74% N_2; 1.23% CO; 11.03% CO_2
23. 7.77% H_2; 35.47% CO; 47.3% CH_4; 9.46% N_2
24. 27.32% CO; 4.14% H_2; 46.1% N_2; 22.44% CH_4
25. 11.53% CO_2; 46.57% O_2; 41.9% N_2
26. 6.78% CO; 4.75% N_2; 85.5% H_2; 2.97% O_2

Chapter 12 (page 236)

1. 2.895 kg
2. 13.2 kg
3. 12.175 kg; 21.25% CO_2; 7.55% H_2O; 71.2% N_2
4. 8.77 kg
5. 15.06:1, 21.1% CO_2; 78.9% N_2
6. by mass: 18.88% CO_2, 2.51% H_2O, 6.28% O_2, 72.33% N_2; by volume: 12.82% CO_2, 4.16% H_2O, 5.86% O_2, 77.16% N_2
7. 11.76 kg; by mass: 19.15% CO_2, 5.3% O_2, 75.55% N_2; by volume: 13.19% CO_2, 5.02% O_2, 81.79% N_2
8. 15.1:1; 86.3%; 2.4 kg

9. (a) 876.2 m^3, (b) 14.06% H_2O; 14.64% CO_2; 71.3% N_2
10. 13.65 kg; 3.08 kg CO_2; 1.08 kg H_2O
11. 386.9 l; 17.48% CO_2; 82.52% N_2
12. 7.325 kg; 46.5% rich
13. 33 390 kJ/kg; 32 760 kJ/kg
14. 11 212 kJ/m^3; 7.86 m^3
15. 37 036 kJ/kg; 35 938 kJ/kg
16. 32 400 kJ/kg
17. 6372 kJ/m^3

Chapter 13 (page 240)

1. 612 W
2. 5.4°C
3. 230 W
4. 83.25 kW
5. 1302 kW/m^2; 2.352 kW/m^2
6. (a) 308.9 W, (b) 43.75 W
7. 105.1 kW/m^2; 7.36°C
8. 9.36 kW; 60%
9. 0.77 W/m K
10. 195p; 118.1p
11. 365.4°C; 33.2 mm
12. 1.92°C/min
13. 123°C
14. 1.34°C/s

Index

Acceleration, 113
  angular, 155
  centripetal, 159
Aneroid barometer, 184
Angle of friction, 82
Angular motion, 153 *et seq.*
  equations of, 156
Atoms, 215
  relative masses, 215

Balancing of co-planar masses, 163
Barometer,
  aneroid, 184
  liquid column, 183
Beckman thermometer, 176
Bomb calorimeter, 229
Bourdon pressure gauge, 186
Bow's notation, 19
Boyle's law, 192

Calorific value of a fuel, 227
Centre of mass, 47
Characteristic gas equation, 195
Charles' law, 194
Coefficient of
  friction, 79
  thermal conductivity, 241
Combustion, 214 *et seq.*
  excess air, 218
  of elements, 216
  minimum air, 217
  products of, 221
Conduction, 240
Conservation of momentum, 144
Convection, 240
Couple, 44

Dalton's law, 195
Differential wheel and axle, 104
Displacement, 112
Distance, 112
Distance–time graph, 114

Efficiency,
  limiting, 95
  of a machine, 94
Engine indicator, 187
Equations of,
  angular velocity, 156
  linear velocity, 118
Equilibrant, 8
Equilibrium, 7
  of concurrent co-planar forces, 9
  of co-planar forces, 35

Factor of safety, 66
Force,
  centrifugal, 160
  centripetal, 160
  definition of, 6
  equilibrant, 8
  moment of, 34
  resolution of, 16
  resultant, 8
  unit of, 2
Frameworks, 21
Friction, 78 *et seq.*
  angle of, 81
  coefficient of, 79
  laws of, 80
Fuels, 214
  calorific value of, 227

Gas calorimeter, 233
Gas thermometer, 178

Index 261

Gases, 192 *et seq.*
  Boyle's law, 193
  Charles' law, 194
  Kinetic theory of, 192
  specific heat, 203
  universal constant, 200

Heat transfer, 240 *et seq.*
Hooke's law, 63

Impact of a fluid jet, 147
Indicator diagram, 190

Joule, definition of, 2

Kilogramme, definition of, 1
Kinetic theory of gases, 192

Laws,
  of friction, 80
  of a machine, 95
  of motion, 135 *et seq.*
Length, definition of, 2
Limiting efficiency, 95
Linear velocity, 112 *et seq.*

Machines, 93 *et seq.*
  efficiency of, 94
  ideal, 93
  law of, 95
  limiting efficiency of, 95
  overhauling of, 97
Manometer, 185
  inclined, 186
Mass,
  definition of, 2
  centre of, 47
  co-planar balancing of, 163
Maxwell diagram, 21
Measurement of
  pressure, 182
  temperature, 175
Mechanical advantage, 93
Metre, definition of, 1
Modulus of elasticity, 63
Mole (or mol), 200
Molecule, 215
  relative masses, 215
Moments, 34 *et seq.*
  of a force, 34
  principle of, 35
  resultant, 35
Momentum, 135
  conservation of, 144

Motion,
  angular, 153 *et seq.*
  equations of angular, 156
  equations of linear, 118
  kinetic equation of, 136
  laws of, 136
  under gravity, 124

Optical pyrometer, 181
Orsat apparatus, 222
Overhauling of a machine, 97

Pin-jointed frameworks, 21
Pressure,
  Bourdon gauge, 186
  measurement of, 182
  partial, 195
  volume diagram, 181
Principle of moments, 35
Pulleys, 100
  Weston differential, 106
Pyrometer,
  optical, 180
  radiation, 181

Radian, 153
Radiation, 241
  pyrometer, 180
  Stefan-Boltzman law, 247
Relative velocity, 127, 168
Resolution of,
  force, 16
  velocity, 124
Resultant force, 8
Resultant moment, 35

Scalar quantity, 4
Screw-jack, 103
Second, definition of, 2
Shear stress, 71
SI Units, 1
Specific heat capacity, 203
Speed, 113
Speed–time graph, 115
Stefan-Boltzman law, 247
Strain, 62
Stress, 61
Strut, 21

Temperature,
  definition of, 3
  measurement of, 175
Tensile strength, 67
Thermal conductivity,
  coefficient of, 241

Thermocouples, 179
Thermometers,
  Beckman, 176
  gas, 178
  mercury in glass, 175
  resistance, 181
Tie, 21
Time, definition of, 3

Universal gas constant, 200

Vector,
  addition, 4
  quantity, 4

Velocity, 113
  angular, 153 *et seq.*
  relative, 127
  resolution of, 124

Weight, 3
Weston differential pulley, 106
Wheel and axle, 98
Work done, 2
  by a torque, 3
  indicator diagram, 190
  pressure–volume diagram, 189